on the
Cutting
EDGE
of
Technology

Cover image (from the
Peter Gabriel video
"Steam") courtesy of
Homer and Associates

Copyright © 1993 by Sams Publishing

All rights reserved. No part of this book shall be reproduced, stored in a retrieval system, or transmitted by any means, electronic, mechanical, photocopying, recording, or otherwise, without written permission from the publisher. No patent liability is assumed with respect to the use of the information contained herein. Although every precaution has been taken in the preparation of this book, the publisher and author assume no responsibility for errors or omissions. Neither is any liability assumed for damages resulting from the use of the information contained herein. For information, address Sams Publishing, 201 W. 103rd Street, Indianapolis, IN 46290.

International Standard Book Number: 0-672-30373-6

Library of Congress Catalog Card Number: 93-84374

96 95 94 93   4 3 2

Interpretation of the printing code: the rightmost double-digit number is the year of the book's printing; the rightmost single-digit, the number of the book's printing. For example, a printing code of 93-1 shows that the first printing of the book occurred in 1993.

Trademarks

All terms mentioned in this book that are known to be trademarks or service marks have been appropriately capitalized. Sams Publishing cannot attest to the accuracy of this information. Use of a term in this book should not be regarded as affecting the validity of any trademark or service mark.

"Willow" ™ and ©Lucasfilm Ltd. (LFL) 1988. All rights reserved.

Composed in Palatino, Futura, and MCPdigital
by Prentice Hall Computer Publishing

Printed in the United States of America

PUBLISHER
Richard Swadley

ASSOCIATE PUBLISHER
Jordan Gold

ACQUISITIONS MANAGER
Stacy Hiquet

DEVELOPMENT AND PRODUCTION EDITOR
Dean Miller

MARKETING MANAGER
Greg Wiegand

SENIOR EDITOR
Tad Ringo

EDITORIAL COORDINATOR
Bill Whitmer

EDITORIAL ASSISTANTS
Molly Carmody, Sharon Cox

COVER DESIGNER
Jay Corpus

DIRECTOR OF PRODUCTION AND MANUFACTURING
Jeff Valler

PRODUCTION MANAGER
Corinne Walls

IMPRINT MANAGER
Kelli Widdifield

BOOK DESIGNER
Michele Laseau

LAYOUT DESIGNER
Juli Pavey

PRODUCTION ANALYST
Mary Beth Wakefield

PROOFREADING/INDEXING COORDINATOR
Joelynn Gifford

GRAPHICS IMAGE SPECIALIST
Dennis Sheehan

PROOFREADERS
Terri Edwards, Mitzi Gianakos, Howard Jones, Linda Koopman,
Wendy Ott, Linda Quigley, Tonya Simpson, DennisWesner

INDEXER
Suzanne Snyder

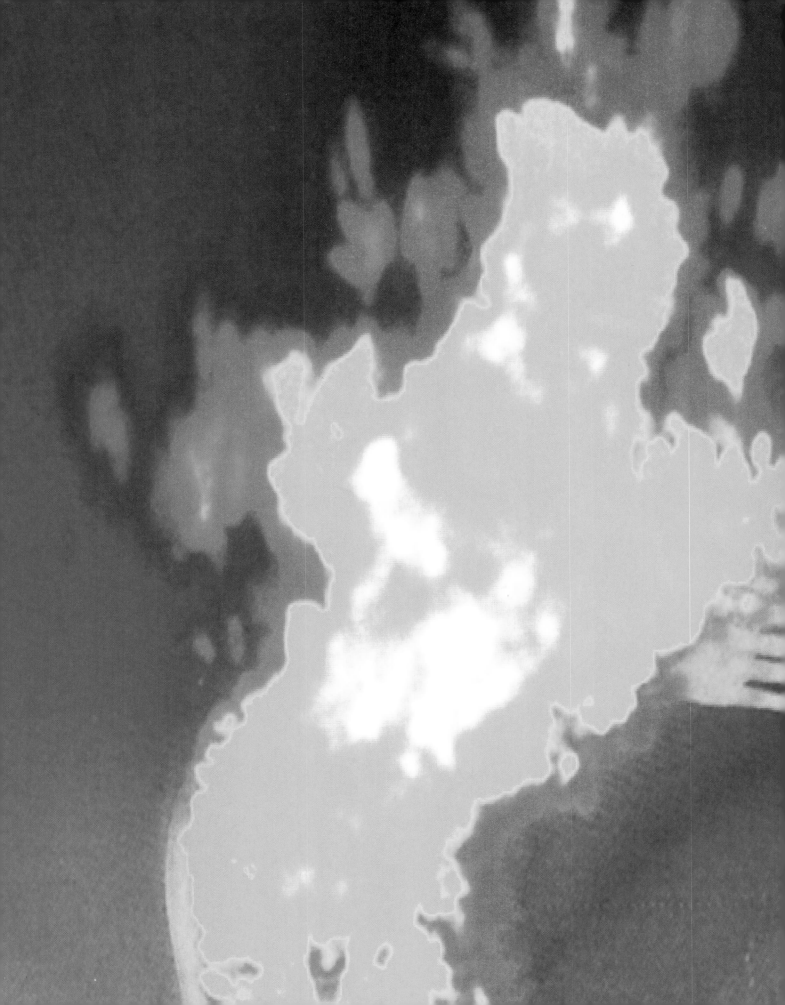

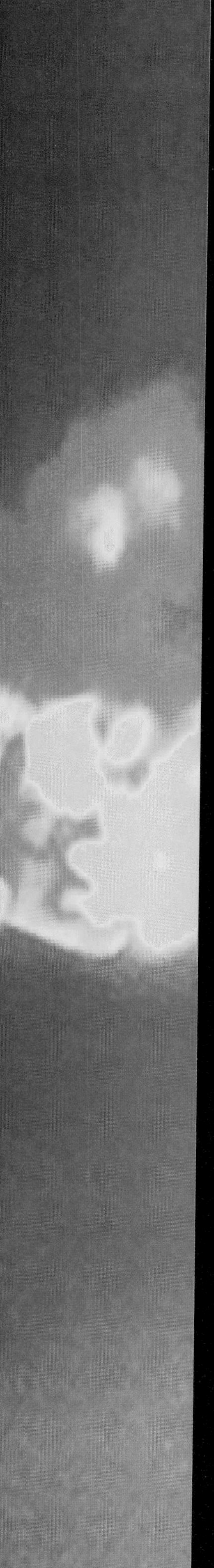

Contents

INTRODUCTION

1 MOTION CAPTURE MEETS 3-D ANIMATION 1

2 INTELLIGENT AGENTS ON THE CUTTING EDGE 15

3 MORPHING 27

4 NANOTECHNOLOGY: PUTTING THE ATOMS WHERE WE WANT THEM 41

5 SMART MATERIALS 49

6 3-D ANIMATION 59

7 WELCOME TO THE VIRTUAL WORLD 69

8 WONDERS OF WATER AND WAVES 81

9 ARTIFICIAL LIFE 93

10 FRACTALS 105

11 FUZZY LOGIC 117

12 INTERACTIVE ENTERTAINMENT 125

13 SOUND ENTERS THE THIRD DIMENSION 137

14 CHAOS AND COMPLEXITY 147

GLOSSARY 159

About the Authors

Dr. Jack Aldridge **Fuzzy Logic** has led and participated in projects in fuzzy control while employed by McDonnell-Douglas Engineering Services and Togai InfraLogic, Inc. in Houston, Texas since 1986. He was a lead participant in projects for NASA's Johnson Space Center, hybridizing fuzzy processing with JSC's CLIPS expert development system and using fuzzy control for heating, ventilation, and air conditioning of buildings and spacecraft environmental systems.

Scott Anderson **Morphing** is the author of *Fantavision*, an animation program with morphing and tweening aimed at kids. Anderson was a contributor to *Tricks of the Graphics Gurus* and the author of *Morphing Magic*, both published by Sams in 1993.

Ivan Amato **Smart Materials** is a staff writer at *Science* magazine.

John Iovine **Artificial Life** is a freelance writer based in New York city. He was written three books, *Homemade Holograms*, *Fantastic Electronics*, and *Kirlian Photography*.

Linda Jacobson **Welcome to the Virtual World** and **Sound Enters the Third Dimension** is a journalist, writer, and speaker who specializes in emerging communications and entertainment technologies. She is the co-author of the book *Cyberarts: Exploring Art & Technology*, and she is currently writing a book titled *Garage Virtual Reality*.

Mike Morrison **3D Animation** and **Interactive Entertainment** is the owner of ddd Graphics, a southern California computer graphics company. Morrison is the author of *The Magic of Image Processing*, published by Sams in 1993.

Gayle Pergamit and Chris Peterson **Nanotechnology: Putting the Atoms Where We Want Them** are co-authors (along with Eric Drexler) of *Unbounding the Future*, a look at nanotechnology.

Barbara Robertson **Motion Capture Meets 3-D Animation** is a senior editor at *Computer Graphics World*. She was the editor and research director for the *Whole Earth Software* catalogue, the bureau chief for *Popular Computing*, and the west coast editor for *Byte*. She has contriubted numerous articles for computer and technology magazines.

William Roetzheim **Chaos and Complexity** is a Senior Associate with the consulting firm of Booz, Allen & Hamilton and a well-respected writer on both management and technology. The author of six popular books on computer-related topics, he lives and works in the San Diego area.

Peter Rothman **Intelligent Agents on the Cutting Edge** is the founder of Avatar Partners, and leads the company's virtual reality related efforts. His expertise includes artificial intelligence, neural networks, non-linear systems and chaos theory, virtual reaity systems, and 3-D computer graphics. Rothman is the host of the virtual reality conference on the WELL.

Peter Sorensen **Wonders of Water and Waves** and **Fractals** is a contributing editor to *Computer Graphics World* and is a computer animation consultant who lives in Santa Monica, California.

Introduction

New technologies surround us, making our lives more simple, more interesting, and more fun. We wear eyeglasses that automatically darken into sunglasses when we step outside on a bright, sunny day. Voice-mail systems make it easy to leave a message for someone. The Space Invaders and Asteroids of old can now teach us about music, art, and history.

There's never enough new technology, and a technology is never new for too long. With every breakthrough, a hundred people might say "Wow," but there's always one who says, "How can I push this a little further?" One inventor's breakthrough is another inventor's stepping stone.

Medicine, biology, aerospace, physics, chemistry, geology, entertainment, and business are just a few of the areas where you can find current cutting-edge technology (from virtual reality to morphing 3-D images). This book examines over a dozen topics on the cutting edge of technology. The authors examine how these breakthroughs affect you and where the technology is leading. Just a few years ago, this information might have been well suited for a science-fiction novel, but today it defines what's new on the scientific front.

Motion Capture Meets 3-D Animation

Animating movement, particularly human movement, can be difficult. Giving an inanimate object (such as a lottery ticket) realistic human movements can be nearly impossible. Not anymore. Now you can capture the movements of a human and map them to a computer-created animation—a technique being used in television commercials and rock videos.

Intelligent Agents on the Cutting Edge

Artificial Intelligence is a term used to describe the use of computers (agents) in such a way that they perform operations analogous to the human abilities of learning and decision-making.

Morphing

Morphing is showing up everywhere, in movies (*Terminator 2: Judgment Day*), music videos (Michael Jackson's *Black or White*), and TV (Exxon commercials). This article looks at how morphing got its start, some scientific uses of the technology, and how ordinary people can perform their own morphing effects on PCs.

Nanotechnology: Putting the Atoms Where We Want Them

Imagine tiny computers so small they must be seen with a microscope. Consider machines smaller than a red blood cell that can circulate through your body and attack and remove infectious organisms. Picture assembling a desired molecule one atom at a time, like a Tinkertoy. All of these notions are ideas behind nanotechnology.

Smart Materials

Bridges, airplanes, the walls in your house—imagine all these constructed materials sensing danger (such as corrosion or fatigue) and averting problems before they occur or changing shapes to prevent destruction.

3-D Animation with Ray Tracing

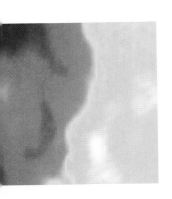

Three-dimensional graphics appear to have height, width, and depth, and lifelike attributes such as texture, shading, and so on. They're usually photorealistic. This section covers the use of 3-D computer animation in commercials and movies and looks at metaballs, a hot new graphics technique used in the movie *Jurassic Park*.

Virtual Reality

Essentially a technique for creating a "simulated experience," virtual reality is an experience of a simulated external world. With the help of computerized goggles, helmets, and gloves, you become part of a digital world.

Liquid Graphics

Making realistic models of waves, water, and other liquids on computers can be one of the trickier assignments in graphics. How does light scatter when it hits the water? What happens to waves? Although the job is difficult, the payoff can be immense, resulting in some breathtaking pieces of art.

Artificial Life

Now you can model and study different facets of life with a home computer. Programs can model evolution, death, reproduction, bird migration, and ant colonies. Some programs even use what they've learned such as programs that evolve to become more efficient.

Fractals

Fractal-art generation is a computer process that creates complex, repetitive, and mathematically defined geometric shapes and patterns that resemble those found in nature. This section introduces you to the fascinating world of fractals and what is involved in creating fractals.

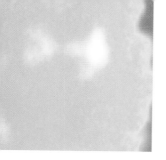

Fuzzy Logic

Not all decisions are "yes" or "no;" some are "maybe" or "it depends." Fuzzy logic looks at complex decisions, and how they can apply to real-life situations. Imagine tossing your laundry into a "fuzzy" washing machine, pushing a button, and leaving the machine to do the rest, from measuring out detergent to choosing a wash temperature.

Interactive Entertainment

Just as new technology can make work an easier and more enjoyable task, technological advances can make play even more fun. Learn what's in store for the future of home video game systems and computer games.

3-D Sound

Three-dimensional sound generally refers to a new type of audio technology that lets recording artists enlarge their musical canvas and gives listeners the opportunity to enjoy the results without buying extra speakers, expensive headphones, or special decoders.

Chaos

Chaos is one of the most fascinating new fields in modern science. Through symmetry, mathematics, and images, Chaos explains order and pattern in nature. Chaos literally means "without form." This section shows chaotic images and explains how they are created.

I hope you enjoy this book. If you have any suggestions of technologies for future editions of this book, please write Sams Publishing at 11711 North College Avenue, Carmel, IN, 46032, or 71450,1704 on CompuServe.

**RICHARD SWADLEY
PUBLISHER
SAMS PUBLISHING**

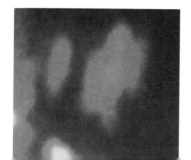

CHAPTER 1

Motion capture meets 3D animation

barbara robertson

In computer graphics, making a ball bounce is not always as easy as it might seem. And making a character move like a human has often seemed nearly downright impossible.

If animators are working in 2-D, perhaps creating a cartoon or an animated film, they have to create drawings (cels) for each position of a bouncing ball. In film, which has 24 frames per second, that means 24 drawings for each second of film. Video requires even more: 30 frames per second. Computer software programs have been created that help animators by calculating, then drawing the motion between key positions, known as key frames. For example, an animator might draw two key frames, the ball at the top of a bounce and when it hits the floor, and leave it to the computer to fill in the rest. Even so, the overall motion sequence has to be planned and scripted by the animator.

cel: a single drawing from an animation.

For animators working in 3-D, the process is similar and more complex. In essence, 3-D animators typically create a wireframe version of a 3-D object or character and position it in 3-D space in several key frames. As with 2-D animation, a software program calculates (interpolates) all the positions between the key frames—a more complex task given the rotations possible in three dimensions. In addition, the animator enters parameters to specify the color and texture of the surface of the model and positions and specifies lights and camera angles—the view we see—for the scene. (The camera might have us look up at the bouncing ball from the floor, look down on it from the ceiling, view it straight on from the side, or move around during the animation.)

Although it might not seem terribly difficult to script the motion of something as common as a bouncing ball, in fact, it's not easy to create smooth, pleasing motion. Even more difficult, and a constant challenge for animators and computer graphics researchers, is scripting the motion of characters. Imagine creating all the individual positions for something as simple as a person walking across a room. As a person walks, her whole body moves—arms, legs, hands, feet, torso, head, perhaps even mouth, eyes, eyebrows. In one frame, the computer character might have her left foot on the floor, right foot in the air, head looking forward, and a smile on her face. In the next frame, the right foot moves a bit, the heel of the left foot begins to leave the floor, and the character's right arm might move a fraction. It's a daunting amount of detail. And it's motion that, because we're all so familiar with it, looks really awkward when it's not done well.

You might wonder, then, why people don't, somehow, simply create 3-D digital versions of puppets. Why, instead of asking artists to painstakingly draw or model each motion, someone doesn't invent a way to control computer models with the digital equivalent of puppet strings?.

Well, that's exactly what's beginning to happen. An elite and slowly expanding group of performance-animation characters, digital puppets if you will, first began arriving on computer screens during the late 1980s. Some companies use actors hidden behind screens to control a character's movements and expressions as well as provide a voice, much as a puppeteer might. Other companies use the techniques more as a movie studio might. At these production studios, directors block out scenes, work with actors, capture motion in several takes, then combine the best motion with a sound track and images carefully rendered "off line" to produce a final film or video. The

motion capture: using current technology to study and save the movements of a human subject, to later apply this data to a computer-generated model.

captured motion helps animators create realistic "humanoid" characters faster and easier than with traditional techniques.

These animators owe the ability to create digital puppets, in part, to one of the most famous puppeteers of all, the late Jim Henson of Muppets fame.

Around 1985 to 1986, Henson approached a computer graphics animation company in Los Angeles called Digital Productions with the idea of creating a digital puppet, according to Brad deGraf, who was then head of technical direction for the Hollywood production company, and who is now with Colossal Pictures in San Francisco. "The imaginative leap from controlling a puppet to controlling a computer-generated character is not a huge one," deGraf says. "You're already taking motion and turning it into something else."

Henson brought with him a "waldo" that had been used to remotely control the puppet frog Kermit. (A "waldo" is the generic name for a control device that mimics the thing it's controlling.) The programmers at Digital Productions managed to use the waldo to control a wireframe computer character, but that character was never used in a production.

At about the same time, in another Los Angeles production studio called Omnibus, Jeff Kleiser was among those trying to capture real human motion for a Marvel Comics pilot. Rather than using a waldo, the Omnibus animators tried a system from Motion Analysis (Santa Rosa, CA) designed for biomedical purposes. This system uses cameras to capture body positions as people move. Although the technology seemed promising, the techniques weren't used for a production.

Even so, these two events set the stage for performance characters to come.

Kleiser left Omnibus to start the Kleiser-Walczak Construction Company with Diana Walczak. The company engaged Frank Vitz, a programmer with some experience with Motion Analysis, and used that company's motion-capture system to animate Dozo, a 3-D computer-graphic rock star who danced in front of a microphone as she sang "Don't Touch Me" in a 3 1/2 minute, computer-animated music video (see Figure 1.1).

Figure 1.1. *Dozo, the first computer-generated rock star to use motion captured from a human dancer, was created at the Kleiser-Walczak Construction Company in 1988. More recently, the same company created state-of-the-art computer animation, some of which uses motion capture techniques, for the Luxor Hotel in Las Vegas.*

waldo: a generic name for a control device that mimics the thing it's controlling.

At about the same time, Brad deGraf formed, with Michael Wahrman, another performance animation pioneer, the deGraf/Wahrman production company. Together, they created a 3-D shaded model that could be manipulated interactively by a person using various types of input devices. "Mike the Talking Head" first entertained people at SIGGRAPH, the popular computer graphics conference sponsored by the Association for Computing Machinery's Special Interest Group on Graphics, during the summer of 1988.

Meanwhile, a company called Pacific Data Images (Sunnyvale and Hollywood, CA) began working on a computer generated character that became Waldo C. Graphic, for Jim Henson.

For Dozo's "Don't Touch Me," music video, Kleiser-Walczak animators used Motion Analysis' optical systems to capture a dancer's motions, then translated that motion into 3-D data.

The motion is captured with video cameras linked to image processors (see Figure 1.2). Special circular or spherical markers made of retroreflective material are attached to key points, such as joints, on a person's body. As the person moves, the cameras—usually three or more—record the points. The image-processing system determines the x and y position of each marker, then calculates (triangulates) the z position. Within minutes, you can see (and replay) a variety of representations of the captured motion from any 2-D view.

Figure 1.2. *This gymnast has markers made of retroreflective material attached to various parts of her body (ankle, knee, waist, shoulder, elbow, wrist, and so on). As she tumbles, her movements can be recorded by six video cameras that each take 200 pictures per second and tracked using the ExpertVision system from Motion Analysis (Santa Rosa, CA).*

To create Dozo, the Kleiser-Walczek animators converted the 3-D data into line segments onto which they draped, in effect, a body with flexible skin. As these underlying segments moved into the various captured positions, Dozo danced. Or, more accurately, the captured motion determined the position of the 3-D Dozo computer model for each frame of the animation.

While Dozo was probably the first computer graphics rock star who danced with captured motion, "Mike the Talking Head," was arguably the first interactive 3-D performance character. Created by deGraf/Wahrman in 1988 at the behest of Silicon Graphics, Mike's face could be manipulated using an input device such as a joystick. His lips, for example, could be made to move as someone talked. The character showed off SGI's new 4-D workstations which allowed shaded images to be moved interactively, as well as the ability of deGraf/Wahrman to create software that could smoothly interpolate the motion between facial expressions that changed whenever someone moved an input device.

And PDI's computer-generated puppet created for Jim Henson and aptly named Waldo C. Graphic, was probably the first TV star to use captured motion. Waldo C. Graphic made his debut on television alongside Kermit. To move the digital puppet: "We built an armature that was sort of like a Luxo lamp," says Graham Walters a Senior Animator at PDI. An animator could move it from side to side, up and down, forward and backward to control the movement of the puppet on the computer screen that was made of about 200 flat-shaded polygons. "It was as if you had your hand inside the (computer generated) puppet," Walters says. After the motion was created, each frame of the 3-D puppet was recomputed at higher resolution with more complex shading (rendering). Then each of those images was composited with another puppet. "It was quite a challenge then," Walters says.

It's still a challenge, but with faster hardware and more sophisticated software, the challenge is increasingly interesting to animators working in production environments.

Homer & Associates (Hollywood, CA), a computer animation company, has been using a visual, camera-based system similar to the Motion Analysis optical system, to create music videos and TV commercials.

In 1992, in association with Harold Friedman Consortium, they produced what they believe was the first TV commercial to use motion capture for animation. Called "Party Hardy," the spot promotes the Pennsylvania Lottery.

In this animation, a crowd of humanoid lottery tickets cavort and gossip at a costume party while they wait for the surprise guest of honor. Each ticket has a different movement, voice, and facial expression that matches his or her costume. Director Michael Kory performed the motion for all the tickets and their faces—including all seven speaking parts.

To capture Kory's motion (see Figures 1.3-1.5), Homer & Associates relied on hardware, software, and services provided by SuperFluo (Los Angeles). SuperFluo, founded by Francesco Chiarini and Umberto Lazzari, began developing applications for 3-D computer character animation in 1988 using the Elite Motion Analyzer produced by Bioengineering Technology and Systems (Milan, Italy). Although there are technical differences, both the Elite and Motion Analysis systems, designed principally for medical applications, have basic similarities. The Elite system also uses video cameras with infrared light sources around the lens, captures the x and y positions of retroreflective markers, and then calculates the z position. However, Superfluo provides custom system software designed specifically to help animators use the captured data.

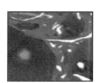

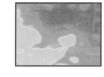

Figures 1.3-1.5. *In this commercial for the Pennsylvania Lottery called "Party Hardy," director Michael Kory of Homer & Associates (Hollywood, CA) captured his own motion for each of the humanoid lottery tickets and their faces.*

To create the rubbery lottery tickets, Kory held up large pieces of foam dotted with markers, and acted each character's part—big swaggering steps, little short steps, and so on—as infrared cameras recorded his movement. The captured data was then used to drive a simple object directly. "It took us three days to capture motion it might have taken three months to create with key frames because there are so many characters," says Kory.

Each character's face began as a clay model sculpted by Kory; the clay model was digitized to create the 3-D computer model. To give the neutral faces on the computer model human expressions, he used computer graphics software to create several individual digital "targets"—a smile, a raised eyebrow, and so on. Once the targets were complete, Kory was ready for motion capture.

Markers were attached to his face ("I looked like I had a weird disease," he says). He calibrated his face to do the expressions he'd built (for example, here's a big smile). And then, the SuperFluo system captured

Kory's facial expressions and lip synch movements as he acted each character's part (see Figure 1.6).

That captured motion was mapped to the predefined target expressions using custom software from SuperFluo; custom software from Homer helped handle the interpolation between the targets. "Each set of targets has the same number of points and the same numerical order of polygons so the software can interpolate between them," explains Conn.

In "Sister of Pain," Conn and Kory used motion capture to animate a 3-D robot clone of the sexy female co-star, the "sister of pain" of the video. The "sister" conjures the robot clone and sends it walking along a laser beam toward the rock band (see Figure 1.7). At the last moment she interrupts the beam, sending the clone falling toward the ground. Luckily, a passing spacecraft whisks her to safety.

The video's choreographer, Susan Scanlon, provided the movement for the robot clone (see Figures 1.8 and 1.9), which was captured with the same Elite video system as Homer & Associates used successfully earlier for "Party Hardy." "The walk the girl does on the tightrope (laser beam) is very realistic," says Conn. Because the clone was a robot, Scanlon could walk along a line in the studio; she didn't actually have to walk a tightrope.

For the scene where the clone falls off the beam, however, Kory had Scanlon jump backwards off a platform onto a gym mat. Remarkably, the entire computer animation

Figure 1.6. *The spots on director Michael Kory's face are made of special light-reflecting material that makes it easier for the Elite system software developed by SuperFluo (Hollywood) to detect motion in images captured with video cameras.*

Figure 1.7. *In this scene from a Vince Neil music video titled "Sister Pain" produced at Homer & Associates for Warner Brothers Records, a computer clone walks the laser beam toward "Sister Pain."*

More recently, the company has been using commercial 3-D software from SoftImage (Montreal, Canada) to channel the motion data between the Elite system and a 3-D model, to create two music videos: "Sister of Pain" for Warner Bros. recording artist Vince Neil of Motley Crue fame, and working for Brad deGraf at Colossal Pictures, Peter Gabriel's video called "Steam."

On the Cutting Edge of Technology

The fire and ice scene from Peter Gabriel's music video "Steam" was created by Homer & Associates, who co-produced the video with Colossal Pictures (San Francisco). Many of the scenes in the video were created with performance animation software from Colossal's Brad deGraf using motion captured from Gabriel and professional dancers.

was completed in three weeks. "It isn't state of the art," says Conn. "We didn't have time to work on facial expression or on the texture of the skin. It was, however, state of the art of what we can do in three weeks."

For "Steam," "We motion-captured Peter Gabriel for two days," says deGraf. "It was two days of the most fun I've ever had at work. Gabriel said, `If you're not working you're dancin.' So everyone in the whole (Homer & Associates) studio was dancing."

Figure 1.8. *In this scene, the "Sister's" computer-generated clone is leaping from a trapeze onto a laser beam.*

Figure 1.9. *An optical motion-capture system from SuperFluo turned this human motion into computer data that moves the computer-generated clone. Here, the human is creating motion for the scene in "Sister Pain" where the clone leaps from a trapeze onto a laser beam.*

Again, they used markers on the dancers and the SuperFluo Elite visual motion capture system to track the motion.

The data drives a stick figure, created with the SoftImage software, with dots on each joint of the 3-D model that match the markers on the dancers.

"It's impossible to get the dots to match exactly because they are on the outside of (the human) body not inside (on the computer model) where we need them," says Kory. "I had to do a lot of hand animation. But although the motion capture is nowhere near automated, it would have been impossible to do the "Sister of Pain" animation by hand. Even with an open-ended deadline, I don't think the best animators could do motion as good."

"With motion capture I get the exact motion I want," he adds.

"It's actually kind of strange. It's almost like I'm really in the computer."

Actually putting actors "in the computer," that is, creating interactive, real-time performance characters and digital puppets has been the primary goal of some animators and some companies.

Brad deGraf, for example, has created performance animation software called "Alive!" which runs on workstations from Silicon Graphics (Mountain View, CA). With "Alive!," an on-screen digital puppet can be controlled by a "puppeteer" using any number of real-time input devices: mouse, joystick, a motion capture system, the keyboard. "For me, the software does two things," he says. "It's great for live performances, and it's orders of magnitude better for creating non-live animation." (See Figure 1.10.)

With Alive!, deGraf can synchronize a character's lips to a soundtrack by moving them with a mouse or a joystick, and save that motion. Then, while that animation plays back, he can add a second layer of motion—

On the Cutting Edge of Technology

Figure 1.10. *Brad deGraf, one of the pioneers of performance animation, created a character named Roscoe Tilda who acted as host for the Spanish pavilion at Expo '92.*

maybe eye movement; then a third layer and so on until he creates an entire animation. "It's less precise than a scripting system," he says, "but it's near real-time."

DeGraf also has given Interactive Personalities (Minneapolis) a license to use Alive! to create performance characters for corporate events. The first character created with the software was "Yoozer Friendly," a talking space shuttle that made its debut at a computer industry trade show (see Figure 1.11). An actor supplied Yoozer's voice and controlled his movements and expressions with a variety of devices including a special hand device with 5 plungers on top of a joystick. The plungers move Yoozer's eyes, eyelids, eyebrows, and mouth, thereby creating expressions ranging from scowls to smiles. Although the company considered attaching motion sensors directly to the actor's face, they decided against it. "We felt we could get as much flexibility with hand control," says Dan Yaman, President, noting that the actor learned how to use the plungers in two days, "and, we can assign the plungers to other things."

The wizards at SimGraphics (South Pasedena, CA), on the other hand, are using motion sensors attached to an actor's face to drive the motion of computer characters they call VActors (virtual actors). A VActor is a computer-generator character or object controlled in real time by people. "We named it VActor because you need an actor to bring the

DeGraf has used the system himself to create a character named Roscoe who acted as host for the Spanish pavilion at Expo 92, and, with Homer & Associates, to help Chris Wedge of Blue Sky Productions create "SpaceBoy in Sky-High Scramble!" (Motion for SpaceBoy was captured via SuperFluo, and previewed with Wavefront software by Homer & Associates.)

character to life," says Steve Glenn, Vice President for New Business Development. "VActors are lifeless unless they're connected to real people."

Using SimGraphics' VActor Performance software, Mario from Nintendo's Super Mario Brothers videogame has become an interactive 3-D character who, along with his evil nemesis Warrio, entertains crowds at consumer electronics and computer trade shows. Mario and Warrio, each a disembodied head, are controlled by actors wearing a face waldo designed by SimGraphics and the Character Shop (Los Angeles) and developed by The Character Shop. The face waldo currently supports 12 inputs, although it could support more according to Glenn. It looks like a cross between something from a science fiction movie and one of the first permanent wave machines used by hairdressers (see Figure 1.12). But it works. As an actor moves his face while he's talking—responding to the audience, perhaps—so does Mario or Warrio. "With a waldo, an actor doesn't have to learn how to puppet," says Glenn.

When first performed, the technology was so new Nintendo had to hire two actors that were the same size to be sure the one face waldo fit both. The second generation waldo is more flexible—and more sensitive.

"When I first used the system, I tried so hard to control every gesture," says Charles Martinet, one of the actors behind Mario and Warrio. "Then, when I realized how well it works, all of a sudden I became the character. I wasn't controlling a puppet, I was the puppet."

The first VActor to have a body as well as a face was actually a Dinosaur named Tarbo created for Fuji Sankei TV. Tarbo served as host for an exhibit called "The Last Dinosaur Kingdom," in an interactive attraction in Tokyo. Tarbo's facial animation is controlled by an actor wearing the face waldo; his body is animated using other input devices such as the flying mouse. The models were created by Viewpoint (Orem, UT). Viewpoint's Steven

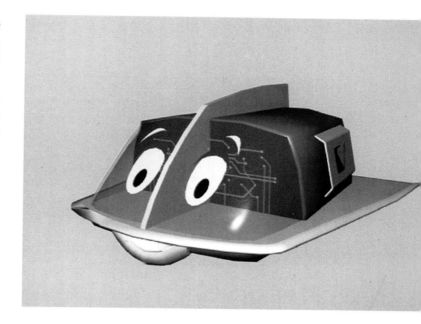

Figure 1.11. *Yoozer Friendly, a computer-generated, talking space shuttle created by Interactive Personalities, worked the crowd (with help from an actor behind the stage) at a trade show.*

Figure 1.12. *Mike Fusco, vice president of engineering for SimGraphics, is wearing a face waldo that captures his head, face, and lip movements and converts them into data. Software developed by SimGraphics translates and transfers that data, in real time, to the character on the computer screen. When Fusco smiles, Tarbo the dinosaur smiles.*

Keele, whose title reads "electronic sculptor," started by making a clay model of the head, then digitizing that model. Once his digital model was approved, he began working on "morph targets" that allow the software to create facial expressions, each a duplicate of a neutral model with the same number of points and polygons, but with some points moved. One target, for example, might have the eyes in a neutral position with the mouth moved; another has a neutral mouth and raised eyebrows. In all, Viewpoint provided more than 50 targets.

Before a performance, the face waldo is calibrated for an actor by matching his hardest smile and biggest puffed cheeks, for example, with Mario's biggest smile or puffed cheek target, or her raised eyebrow with Tarbo's.

During a performance, SimGraphics' software has to recognize the changing expressions being captured by the face waldo in real time, then combine bits and pieces of its targets to match the expressions—on an ongoing basis, smoothly. "The software might use 1/10th of one target, and half of another," says Mike Fusco, vice president of software. "We do blends a lot."

Until recently, the emphasis at SimGraphics has been strictly on performance animation. Now, however, the company is planning to offer VActor/Production, which allows production studios to create real-time animations that can be improved visually *offline* by rendering each frame at higher resolutions with more texture, or perhaps by compositing the computer character with live action or other objects in a 3-D scene.

Real-time capture and display techniques are also being used in production studios such as PDI, which is now working on its third generation "waldo," a full-body suit. "What we like is that we can put a performer in the suit and let them be spontaneous, improvisational," says Carl Rosendahl, President.

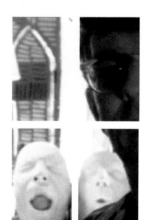

At Mr. Film (Venice, CA), Chris Walker is also using a custom designed full-body suit and Ascension motion sensors along with other technology to choreograph computer animated characters.

"For us, the motion capture is a foundation for the animators to use," says Chris Walker. "It's like having a stage set with real characters. The director can say, `I like this.' It lets the animators do the highest 40 percent."

Some animators, anyway.

"I don't think [motion capture] is the ultimate solution for moving every character," says Rosendahl. "It lends itself well for very human characters or puppetted characters. I love Kermit and I love Mickey Mouse, but they have different movements." (See Figure 1.13.)

Figure 1.13. *In a pilot for the Discovery Channel's "The Next Step" TV show, a computer character acted as co-host with Richard Hart. The character was "puppeted" by an animator at Pacific Data Images (Sunnyvale and Hollywood, CA).*

Walters agrees. "I don't see it as better or worse. Instead, it's a new art form."

In addition, Rosendahl is quick to caution people not to have false expectations about producing quick, low-cost, broadcast- or film-quality computer animations simply because motion can be captured. "Anyone telling you they can do a half-second animation in a half second is only telling you that they haven't done it yet," he says. "It takes a long time to do a minute of performance."

Homer & Associates' Kory agrees: "I used to want people to think we could do an animated TV series because of motion capture. But we're still a long way from that. At this stage, it still requires a lot of tweaking.

However, the technology is improving. Workstations, particularly those from Silicon Graphics' are making it possible to work interactively with increasingly sophisticated, rendered 3-D models. Software companies such as SoftImage and SimGraphics are developing ways for animators to use motion capture without having to write their own software. And the promise of virtual reality is driving more and better motion capture, sensing, and tracking devices into the market at lower prices.

SoftImage, for example, frequently demonstrates at trade shows how magnetic motion capture systems such as the Polhemus' FastTrack device can be used to create real-time, full-body animations. Magnetic devices, used extensively in the Virtual Reality field to track the motion of head mounted displays, have one or more emitters that emit a magnetic field and many receivers (sensors). The emitters and sensors are linked with cables to one or more processing units that can track the position of the receivers in the magnetic field as someone moves. In their demonstrations, a mime dressed in black, wired to the motion tracking device, dances while his or her doppelganger, a figure on the computer screen, matches every move. Using the "channel" feature in SoftImage software, the motion tracking devices attached to the mime are assigned to particular points in the wireframe character. "This is the future of computer animation," says Morin (see Figure 1.14).

So, what can we expect in this future? For one, more characters driven by motion control and more computer-generated real-time, interactive characters—perhaps even hosting TV shows.

For another, the use of motion captured from humans to animated non-humanoid objects.

For example, this year, in the atrium of a new hotel (the Luxor) to be built in Las Vegas, a theme park created by Doug Trumbull of movie special effects fame, features three films that include computer graphics animations by Kleiser-Walczak. One of the films, a 3-D stereo movie, was created entirely with computer animation and uses captured motion throughout. Kleiser-Walczak once again used the visual systems from Motion Analysis to drive the animations, and again worked with Frank Vitz. "We didn't want any encumbrances on the dancer," explains Jeff Kleiser. "We wanted it to be very improvisational."

In this film, the captured motion is used not only to drive computer characters, but also to create the motion for DNA coils, a galaxy, and water.

"We started with a sound track," says Kleiser, "and had two dancers with `witness points' (markers) moving to the sound." Then they applied the motion to non-human objects in the film.

"What we're trying to get across is the idea that there is a life force that permeates everything," he says, "so with human motion applied to objects it implies visually that galaxies are alive, that everything is part of a universal life force."

In the future, the use of motion capture for animation might also evolve into systems used in fields other than the entertainment industry.

A pediatrics unit of the Loma Linda University Medical Center (Loma Linda, CA) hopes to install a VActor/Performer system to produce real-time characters in a new wing of the hospital. "Children will be able to tune to a special channel

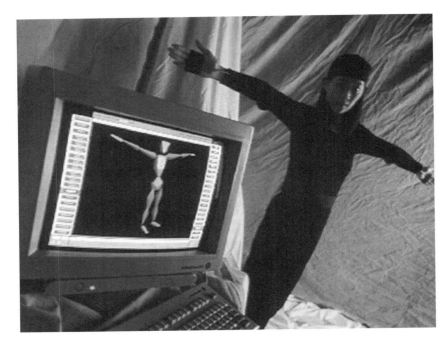

Figure 1.14. *SoftImage (Montreal, Quebec, Canada) was one of the first software companies to add methods for feeding data from motion-capture devices into its 3-D animation programs.*

and a character will come on," says Dave Warner, a medical neuroscientist in the Center's Neurology Research Center. Warner has already worked with SimGraphics and actor Charles Martinet to produce a test run of such a system. He expects the characters, who will probably be acted by therapists, can reinforce doctors' orders, explain operations, and help with pediatric psychology. "Who knows what a child might tell a teddy bear," he says. Or a fuzzy star, or a rainbow.

But it's more fun to imagine the far future. In the far future, using new generations of this equipment, children might be able to create the teddy bear characters themselves. They'll step into a body suit, act out a part, then take home a videotape of their personal animated cartoon. Why not?

(This chapter is based on a article that appeared in *Computer Graphics World*, October, 1992.)

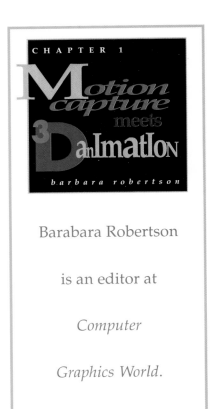

Barabara Robertson

is an editor at

Computer

Graphics World.

CHAPTER 2

INTELLIGENT Agents On the Cutting EDGE

peter rothman

Just Another Day in Cyberspace...

Midterms were approaching and Ann knew it was time to crack the books. She rolled her wheelchair up to the desk, put on her virtual reality glasses, and booted up her cyberspace deck. Ann found herself in her work area, a bare room in which three dimensional icons floated. She pointed at one of the icons with a wand that sat on her desk. A small animated dog appeared. "Good morning Ann," said Odie. Odie was Ann's personal assistant, which she had configured to appear as a friendly cartoon dog.

Odie took Ann's messages, filtered her mail for junk, and reminded Ann of important appointments. "You have an e-mail message from John. Do you want to see it now?" asked Odie. "No, I need to go to the library now," replied Ann. "OK. I'll save the message for you. What do you want to do now Ann?" said Odie. Ann replied, "Odie take me to the library please."

Ann's work area disappeared as Odie transported Ann through the network to the university library. At the library Ann was greeted by the Librarian, a bookish-appearing older gentleman with horn-rimmed glasses. "Good morning Ann, how can I help you?" Ann said, "I need to find information on Christopher Columbus for a report." After a few minutes, the Librarian responded with a list of books, articles, and manuscripts related to Columbus from libraries located around the world.

After reading some articles, Ann became curious about the preparations required to make such a long and dangerous voyage into the unknown. "I'm curious about the supplies that Columbus took on his voyage," Ann said to the Librarian. "OK. I'm searching..." responded the Librarian. While Ann was waiting, the Librarian peered into models of libraries and other document archives around the world.

After several hours of perusing innumerable books and other documents, Odie appeared again. "Ann, you have another e-mail message from John. Do you want to see it now?" Ann thought that it was about time for a break from the research so she said OK, and John's message appeared on a floating slate off to her left. It was just another day in Cyberspace.

What are agents?

The scenario just described might sound far-fetched, but in fact, there are numerous ongoing projects which are attempting to implement the technologies alluded to. Today, most of the technology required to create intelligent agents that lead users through virtual worlds exists. This chapter describes some of the intelligent agents that have been built, the technologies used to create these agents, and some of the possible implications of agent technology.

According to Brenda Laurel, one of the leading proponents and developers of agent technology, "an agent can be defined as a character, enacted by the computer, who acts on behalf of the user in a virtual (computer-based) environment." But just what are intelligent agents? And how are they built? This chapter attempts to answer these questions.

From a technology standpoint, intelligent agents are specialized software tools that are employed to perform tasks for users in virtual reality. Stated another way, intelligent agents are the entities which reside in virtual environments. Agents can perform actions for users of a virtual reality system, they can compete against human opponents in games, and perform many other important functions.

Typically, an agent is implemented by coupling an artificial intelligence program to a graphical representation of a character or object which represents the function of the program. For example, an intelligent database-searching program might be represented by Sherlock Holmes, a science advisor by Einstein, and so on.

The structure of an agent can be broken down into four levels. These levels are listed in the following table in order of increasing levels of complexity.

> **intelligent agents:** specialized software tools that are employed to perform tasks for users in virtual reality.

Table 2.1. Agent Architecture Levels.

Level Name	Description
1. Rendering Level	Performs the graphical, sonic, and tactile rendering of the agent in the virtual world.
2. Dynamics/Kinematics Level	Controls the motion of the agent and the objects comprising the agent in space and time. Typically Newtonian physics is used to define the dynamics.
3. Instruction Level	Actions of the agent are controlled by statements in a formal "programming" language.
4. Planning Level	The agent creates "plans" based on "models" of itself and the environment from which action sequences consisting of instruction level commands are generated.

There exists a continuum of levels of autonomy ranging from virtual puppets, which are completely controlled by a human operator, to fully autonomous agents, which do not employ human input in their decision processes. Moreover, a single agent can operate at different levels of autonomy depending on the situation.

The level of autonomy depends partially on the degree and type of coupling that exists between the agent, the human users, and the simulated environment. At one extreme, the agent is tightly coupled to a human controller, and the agent's actions depend almost entirely on direct inputs from the user. An example of this type of agent is Mario, an animated character controlled by a human operator wearing a face waldo (see Figure 2.1).

Agents that operate essentially independently from human control and intervention are autonomous agents. Rodney Brook's robot Genghis is an example of an autonomous agent (see Figure 2.2).

Applications of Agents

The following sections describe some of the possible applications of agent technology. Some of these applications are currently under development by a variety of researchers, while others remain in the realm of science fiction.

> **virtual puppet:** an intellegence agent completely controlled by a human operator.

On the Cutting Edge of Technology

Figure 2.1.
Mario and the Simgraphics Face Waldo.

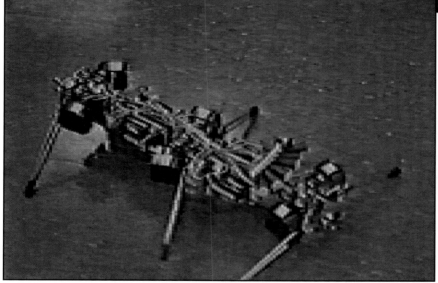

Figure 2.2.
Genghis: an autonomous agent.

GAME-PLAYING AGENTS

Agents that employ strategies, adaptation, or other means to engage a human player in a challenging game of chance or strategy, or to play a player's "role" while the user is not engaged in playing the game. Many game-playing agents already exist. One of the best-known game-playing agents is the "chess machine," a small device that incorporates artificial intelligence, graph search, pattern recognition, and other techniques to play chess. Chess machines today not only play a credible game of chess, but in fact can play at the master level, and are capable of beating most human players.

Many other sorts of games employ agents. For example, one of the "classic" computer games for the Apple II computer, *The Ancient Art of War*, allowed the user to select from a list of famous military commanders for an opponent. Each opponent had a different personality and preferred certain tactics. A more recent example is the latest in the Ultima series of fantasy games, Ultima VII. In Ultima VII, the player is free to explore the entire world of Britannia, meeting many different characters that she can converse with.

> Game-playing agents also have applications in economic analyses and military simulation. Perhaps the best known of these sort of agents were developed at the Rand Corporation in Santa Monica California. Known as Ivan and Sam, these agents were used to examine strategic problems during the Cold War. Ivan and Sam are particularly interesting because they are **endomorphic anticipatory agents**, a mouthful which means simply that the agents use models of each other and themselves to make predictions about the future state of the world.

CYBER-SECRETARIES AND PERSONAL ASSISTANTS

An agent that performs common secretarial tasks such as answering phones, taking messages, receiving faxes, filing, and so on. Many aspects of the cyber-secretary already exist to some degree in current voice mail and Fax back systems. With voice mail, the user is stepped through a series of menu choices by a computerized agent that provides audio instructions. Fax back provides the system with the capability to Fax product information or other images to the user in response to a user's interaction with the system. Products exist which allow voice mail and Fax back to be controlled by a rule-based expert system allowing for a much wider variety of responses than current voice mail systems provide.

> endomorphic anticipatory agents: agents that model themselves and other agents to make predictions about the future state of the world.

Video phones will lead to the use of recorded video images as well as voice interaction with users of these advanced phone systems. Alternatively, the user will interact with a completely synthetic animated character generated entirely from computer models. In the not-too-distant future, all of these technologies will merge with advances in speech-recognition technology to enable the user to have limited spoken interactions with intelligent agents controlling voice mail, Fax back, and other systems.

RECORDED PERSONALITIES AND VIRTUAL ACTORS

Digital models of famous, interesting, or just about any other person can be created using the technology of artificial agents. These models will mimic the person's appearance, particular mannerisms, and potentially be able to respond to simple command scripts. Nadia and Daniel Thalmann have made a short animated film using computer-generated three-dimensional animations of Marilyn Monroe and Humphrey Bogart (Figures 2.3 and 2.4). In the future, the images of other famous stars might also be recorded and used to generate intelligent agents that can emulate performances of the real person.

Of course virtual actors need not only represent real people, they also can be fictional entities such as Mickey Mouse, Super Mario, or the Pillsbury Doughboy. These virtual characters are already popular in advertising and entertainment.

On the Cutting Edge of Technology

Figures 2.3 and 2.4.
Humphrey and Marilyn.

The effort to create realistic computer-generated characters that exhibit human-like emotional characteristics is being pursued by Joseph Bates, a computer scientist at Carnegie Melon University. At CMU, the Oz project has created an animated environment in which four "blobs" interact with each other in a simple virtual world (see Figure 2.5).

INFORMATION RETRIEVAL AND ANALYSIS

Agents can act as information ferrets, searching for data in large databases. These information-seeking agents will be given graphical appearances and they will explore three-dimensional worlds of databases. Advanced capabilities based on pattern recognition techniques include automatic detection of patterns in data, change detection, or transaction tracing. Because these agents might be able to trace or deduce private data from publicly available data there are extensive privacy concerns with these sorts of agents.

Perhaps the best known project utilizing intelligent-agent technology to access a database was the Grollier Guides project developed as a demonstration by Apple Computer. The Guides project was a multimedia system that provided an agent-based interface to a hypermedia database of American history during the period from 1800 to 1850.

The Guides project explored the use of agents representing characters from the historical period contained in the hypermedia database. These agents used a narrative approach to lead the user to explore the database from multiple perspectives.

Figure 2.5. *Bate's Blobs*.

In a recent movie, *Until the End of World*, director Wim Wenders included a scene in which an animated bear searches global databases in order to locate individuals by their credit transactions.

By combining these advanced database systems with pattern recognition and machine learning technologies, it is possible to create agents that can acquire knowledge and discover patterns in data. These agents can infer relationships not contained explicitly in the database. While these agents have incredible potential to aid scientific discovery, they also have the potential for abuse. A database inference agent might be used to infer private information about an individual from public records for example.

Agents also can be used for change detection or alerting the user of the occurrence of an important or interesting event. For example, in

Although the Bounty Bear is entirely fictional, the basic technology depicted is very real. The now infamous PROMIS case management system developed for the Justice Department can be used to track criminals and defense attorneys based on their prior records in federal court cases. Another example of a similar system is the Information America on-line computer database which cross indexes the Postal Services National Change of Address (NCOA) file, subscription lists, birth records, driver's license records, telephone books, voter registration records, and records from numerous government agencies.

On the Cutting Edge of Technology

Figure 2.6.
JACK Agents.

an application developed to aid traders visualize stock market data, an agent known as LIA (for Limited Intelligence Agent) was developed to notify the user of insider trading, large price or volume moves, or other events.

Generation of complex animation sequences

Agent techniques can be used to generate complex scenes containing many moving objects. In the standard key framing approach to generating animations, a sequence of frames is generated that depict the characters in different positions. The "in between" frames are then generated automatically from the key frames through a process of interpolation

In contrast, using intelligent-agent technology, the animator is able to interact with the animated characters through the use of English language commands or by defining the agent's actions and goals in a formal programming language. Badler has developed a system called JACK, which can be used to generate narrated animations containing human-like animated figures. JACK provides a natural language interface to the animation system which allows the animator to control agents with simple English sentences and scripts. Using JACK, the animator can instruct a character to walk across a room and turn on the light switch instead of crafting a sequence of key frames by hand.

Agents also have been employed to provide interesting backgrounds for conventionally generated animations because they can easily be made to interact and behave in unpredictable ways. With the current state of the art, animators can use virtual reality and agent technologies to lay out animation that will be later rendered using high-resolution rendering engines such as Autodesk's 3-D Studio or Pixar's Renderman. As computer hardware increases in power, it will become possible to generate commercial-quality animation sequences employing agents in real-time.

AGENTS AND DEEP INTERACTIVE WORLDS

The ability of a user of a virtual reality system to suspend disbelief is dependent on the *degree of immersion* in the environment and the *level of interaction*. The degree of immersion is a function of the quality of the visual, aural, and tactile rendering as well as the lags inherent in the tracking and display system. The level of interaction is a function of the simulation and logic underlying the behavior of the objects comprising the virtual world.

When the user of a virtual world attempts to interact with an object in a way which is not supported by the simulation, the user's suspension of disbelief is often shattered. For example, the user attempts to operate the stove in a virtual kitchen with no effect. Agent technology provides a means to construct virtual worlds with a deep level of interaction. In these worlds, the users will be able to interact with a wide variety of objects, mechanisms, and agents which exhibit complex and realistic behaviors.

Imagine trying to construct and populate a virtual New York city in which a complete three-dimensional database of Manhattan is used as a backdrop. How can the virtual world designer create characters, places, and objects for the user to interact with without being overcome by the sheer complexity of constructing these objects? Agent technology provides the answer.

Individual characters can be generated, each of which has its own goals, behaviors, and needs. These agents are allowed to interact to create a computational ecosystem or artificial economy in which the agents interact to increase their personal chances of survival or wealth. The behavior generated by such systems of interacting agents can be extremely complex, and with the some effort can be made to approximate the behavior of the natural systems they model.

> **computational ecosystem:** Also referred to as an artificial economy, a system in which artificial agents compete to increase their personal chances of survival or wealth.

BUILDING INTELLIGENT AGENTS

As described earlier, an intelligent agent consists of an artifical-intelligence program coupled to a graphical representation of the agent. This section discusses some of the technologies that can be used to construct the programs that control an agent's behavior as well as its appearance.

Proxies, legal representation, and investment management. This category of agent includes a variety of roles in which the agent is given the legal authority over an individual's property or is allowed to exercise certain legal actions such as initiating or responding to lawsuits. A good legal proxy agent might be employed by firms to avoid litigation, manage retirement fund portfolios, respond to common business correspondences, and so on.

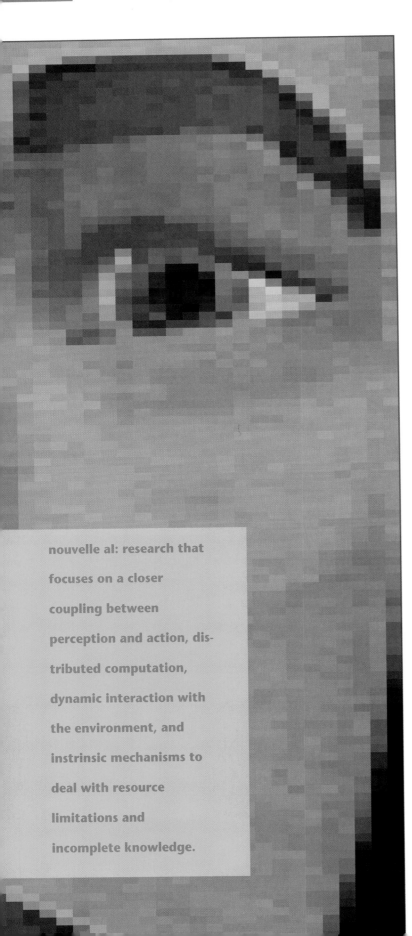

Perhaps the most common approach to controlling agent behavior is to use techniques originally developed by classical artificial intelligence researchers. These include rule-based expert systems, planning systems, model-based reasoning, and other techniques.

The most important of these are planning systems. A planning system is simply a problem solving system which explores potential solutions to a problem *before* executing the steps comprising the solutions. A plan is the sequence of *actions* which produces a solution to the problem of interest. Most planning systems generate several alterantive plans which are examined using heuristic rules to determine which of the plans is most likely to produce a solution or produces a solution with minimum cost.

Two of the central issues in planning systems include planning under uncertainty and time-constrained planning. Because an agent generally does not have access to a completely accurate model of its environment, the planning system must be able to deal with uncertain, incomplete, and even inconsistent information. Moreover, the agent must be able to recover gracefully from situations in which a plan fails as a result of uncertainty about the state of the world or the expected results of the agents actions.

Time-constrained planning addresses the issues of generating plans in real-time. A time-constrained planning system reasons about how much time it has to generate a workable plan before begining the planning process. If a satisfactory plan is not generated within this time period, the planning system can use the best plan found so far or a default plan which is determined *a priori*.

Although classical artificial intelligence approaches are valuable, systems based on these approaches are not able to operate successfully in unconstrained environments in real-time. As a result, an alternative approach known as nouvelle AI has been proposed by several researchers including Brooks, Maes, and others. Nouvelle AI research focuses on a closer coupling between perception and action, distributed computation, dynamic interaction

> nouvelle al: research that focuses on a closer coupling between perception and action, distributed computation, dynamic interaction with the environment, and instrinsic mechanisms to deal with resource limitations and incomplete knowledge.

with the environment, and instrinsic mechanisms to deal with resource limitations and incomplete knowledge. In the nouvelle AI approach, knowledge about the world is not represented in symbolic form as it is in the classical AI approach. Instead, nouvelle AI suggests that the environment is its own best model.

One of the most exciting areas of agent research is the development of agents that learn and evolve. Many different approaches to developing agents that explore and learn about their environments have been developed. Perhaps the two most interesting of these learning technologies are artificial neural networks and artificial life systems.

Artificial neural networks provide another technology with which to build agents that can learn. Artificial neural networks are connected systems of simple processing units whose behavior is determined by the existence and strength of the interconnections between the units. These systems are modeled on the current understanding of neural processes in humans and animals, although the neural networks commonly employed in applications have been developed that can recognize objects, read handwriting, and perform many other functions that are useful components of an intelligent agent.

Artificial life is a relatively new field of study which is attempting to understand the basic principles of living organisms and ecosystems through the use of computational models and systems that exhibit these behaviors. A-life programs typically utilize a genetic algorithm to allow agents to evolve and develop new behaviors in response to environmental conditions or other factors. Genetic algorithms are computer programs that operate on the same basic principles as natural evolution, including sexual reproduction of offspring, random mutations, and natural selection.

In addition to the program controlling an agent's behavior, a three-dimensional model of the agent must be constructed. Depending on the type of agent, this model can be built with a Computer Aided Design (CAD) program or by digitizing a plaster cast of the character.

AGENTS ON THE CUTTING EDGE: THE FUTURE OF AGENT TECHNOLOGY

The current state of the art in intelligent-agent technology is very limited. Existing agents are primarily separate systems which are not integrated into virtual worlds. Significant effort will be required to "bind" these separate agents to virtual worlds. Moreover, many of the existing artificial intelligence programs that will be used to construct agents do not operate in real-time. Real-time execution of these programs will be required in order for agents to support real-time user interaction. Significant advances in the state of the art in a variety of computer technologies will be required to make real-time intelligent agents commonplace.

The ability to interact with intelligent agents as a means of accessing networks of databases and information systems has the potential to radically empower those individuals who have access to the necessary technology. Agents will reduce the amount of time spent doing busy work, responding to junk mail, and so on, freeing individuals to perform more creative and rewarding tasks. Agent technology will enable the construction of new and deeply involving worlds for entertainment purposes. These worlds will combine education and entertainment in a single rich medium, allowing exploratory and experimental learning and encouraging creative thinking.

Agent technology undoubtedly has some potential for very negative impacts on society also. The use of agent technology as a replacement for human beings might lead to widespread unemployment particularly in settings where the user is already interacting with machines. Examples include bank tellers, telephone operators, receptionists, and others whose jobs can be readily transcribed into simple sets of rules or typical cases.

Database inference systems have the potential to allow for widespread and devastating invasions of personal privacy. These systems also can be used to infer classified or proprietary information from public-domain knowledge, possibly encouraging new forms of international and industrial espionage.

Finally, it is not clear what effect prolonged and repeated interaction with non-human agents will have on people. Will it lead to a general dehumanization of the individual? Or will people come to have a greater appreciation for those characteristics that set people apart from machines? Only by considering these issues in the design, implementation, and use of agent-based systems can we be assured that the positive, empowering aspects of the technology will overcome the negative aspects.

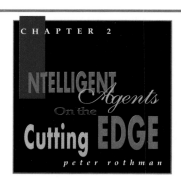

CHAPTER 2

Intelligent Agents On the Cutting EDGE

peter rothman

Peter Rothman is currently writing a book on Intelligent Agents, to be published by Sams in the fall of 1993.

CHAPTER 3

Morphing

scott anderson

Everywhere I turn these days, it seems people are talking about something called morphing, which they saw in *Terminator 2*, a Michael Jackson video, or some shaving commercial. What *is* morphing and what does it have to do with movies, rock videos, and commercials? And why in the world would teachers, students, artists, marketing people and researchers be interested in such a thing?

Morphing is a merciful shorthand for metamorphosing. The word is stolen from biology, where it describes nature's wonderful magic trick of turning a caterpillar into a butterfly or a tadpole into a frog. Morphing depicts objects that smoothly change shape, such as Michael J. Fox becoming a werewolf or Michael Jackson turning into a panther.

> morphing: Shorthand for metamorphosing; it refers to a sequence of smooth changes between two different images, such as a man and a werewolf, or a bat and a vampire. Technically, morphing combines a *warp* and a *dissolve*.

It turns out that the roots of morphing go deep, invading the turfs of both science and art. Mother Nature loves a good metamorphosis, and has many up her sleeve. Educators need the tools to illustrate those transformations. Artists want to automate animation, and morphing is a key to making their lives easier. Movie directors crave the latest morphing effect—at any cost. Even auto makers are using morphing to show you how this year's model has changed. Everyone is morphing!

A Little History

One of the first animators to use film was Emile Cohl, in France. Between 1908 and 1918, the prolific Cohl produced more than 200 animated shorts. A few frames from one of his earlier films is shown in Figure 3.1. Wouldn't you know that among the first filmed animations there would be a morph?

Cohl's style was dreamy, with an emphasis on metamorphosing shapes. He must be credited as one of the first morphing magicians.

Throughout the twentieth century, morphing has been an important tool for the cinematographer. Besides the obvious special effect—Wolfman, Dracula, and so on—there is a milder use for it. It makes an effective cut: fade out on one subject, fade in on another. Done effectively, there is a curious, mystical quality to it.

But the real draw is the magical metamorphosis—an impossible transformation, realized on screen. The effect is so hypnotizing that it was worth spending days to get a few seconds of film.

The Origins of Computer Morphing

In just a few short years, computerized special effects, including tweening (short for in-betweening) and morphing, have made enormous inroads into the Hollywood establishment. These techniques have put new vigor into the cartoons and an amazing new "realism" into feature films.

For the Wolfman transformations, they would strap down the hapless actor (usually Lon Chaney Jr.), and immobilize his head. For the next few hours, the makeup man would apply a few hairs, then the cameraman would shoot a few frames. A little putty here and there, a few more hairs, and shoot more frames. It was agony for everyone involved, it was easy to screw up, and it was fantastically popular.

Figure 3.1.
An early animation, circa 1910, shows a house metamorphosing into a face. And you thought morphing was new.

How did the modern form of morphing arise? Was it just artistic experimentation? That would make a nice story, but the truth is somewhat more prosaic. It turns out that morphing and warping were important to scientists before the artists got hold of it.

One of the first uses of warping was by NASA in the 60s. As they started to compile the satellite pictures of the earth, they ran into a problem. The pictures didn't overlap properly, so it was hard to assemble a large mosaic. The camera lenses introduced some distortion or warping, the angles of each shot were different, and some of the pictures were taken at widely different times. It was your basic mess.

But by finding some common crossroads and other landmarks, they were able to warp the image to correct the distortions. After that, it was a simple matter to stitch the images together.

NASA found another use for warping when it got back pictures of other planets. Some of the spacecraft had unusual optics to perform extraordinary tasks in tight spots. They took strange pictures that had to be warped back to normalcy. The warping function used was the opposite of the warp introduced by the lens, so it was really "unwarping" the images. Most of the pictures of Mars, for instance, were unwarped before you ever saw them.

So warping and morphing started out as serious scientific tools. And then an artist saw it.

In the early 80s, Tom Brigham started to experiment with morphing as an art tool. He was at the New York Institute of the Arts when he started to apply these scientific algorithms to an art project. For his first film trick, Tom turned a woman into a lynx.

It was premiered at the 1982 SIGGRAPH (Special Interest Group on Computer Graphics of the Association for Computing Machinery) convention. It has been described as a hallucination come to life.

Fame didn't rush up to meet Brigham. It took five years before Doug Smythe at Industrial Light and Magic (ILM) took another look at Brigham's concept. ILM is a division of LucasArts, set up by George Lucas to do the special effects for the movie industry.

---

tween: this is a somewhat silly-sounding shorthand for in-be*tween*ing. It refers to a series of images that smoothly connect two other images. If the images are similar—say a horse in two different poses—the result is regular animation. If the images are different, tweening creates a smooth metamorphosis called morphing.

warp: a technique for stretching and squashing an image.

Another remarkable use of warping is in the medical field. A technique known as digital subtraction angiography (a typical medical mouthful) uses two X-rays of a patient. One is taken before and one after the injection of a dye. The dye is opaque to X-rays, so it shows up black on the pictures. To eliminate extraneous clutter, like bones and organs, the first picture is subtracted from the second. That gets rid of everything except the dyed arteries.

It is a great technique, but it only works when the patient is completely motionless. Hopefully, the patient is still breathing. Unfortunately, that messes things up—unless, of course, you can warp the images into registration. Then you get some spectacular pictures of the patient's arteries.

Smythe needed a way to metamorphose several characters for Ron Howard's movie *Willow*. His clever solution is discussed later. But for now it is worth noting that Smythe is the person who coined the term "morphing." The rest is movie history.

In 1993, Brigham won an Oscar for his wonderful contribution to cinema. It was the first time the Academy of Motion Pictures awarded a prize to an algorithm. It probably won't be the last.

The Secrets of Morphing

There are two parts to the morphing algorithm: the warp and the dissolve. The dissolve is (theoretically) simple, but there are a thousand ways to warp.

Warping distorts the main outlines of the two images. The idea is to stretch and squeeze parts of one picture (such as the eyebrows) so that they match up with the equivalent parts of the target picture.

Imagine that you have a map of the world printed on a sheet of rubber. What would happen if you grabbed some part of the picture, say New York, and pulled it south? The question is actually something that computers can address in a simple way.

The screen is a map of colored pixels, and they can be manipulated like any other mathematical coordinates. For instance, to slide the picture to the right, you could simply add 1 to the x-coordinate of all the pixels, and then replot them. To rotate, squash, or skew the pixels, some standard equations will do the trick.

In short, a whole host of mathematical incantations can be brought to bear on the poor pixel, to shove it anywhere you want, on or off the screen.

This process is fortuitously called "mapping" by the mathematicians. If the image stays connected through the transformation (no rips), it is called warping. In most warping routines, the warping is local, so if you pull New York south, Newark moves with it, but Chicago barely budges.

The simplest method is to indicate points on the screen, and then drag them where desired. Although relatively simple to program, this method poses difficulties for the user. The morphing artist must provide hundreds of points for effective warping and keep track of all of them. If even one point is misplaced, the warp will be wrong.

A better method is to use lines. With just two end-points, a line automatically specifies a string of points in between. Another benefit of this implementation is not as obvious: with lines, you can specify global changes. With one line, you can rotate or scale the entire image. This technique is much more powerful than local warping.

The line method was pioneered at PDI and was used in the Michael Jackson video *Black or White*.

Warping has a long and varied history. A house of mirrors is a good metaphor for mathematical warping. In fact, any curvature at all in a mirror causes warping. A convex mirror creates a fish-eye effect, where close objects are normal size, but far objects appear smaller.

Painters since the eighteenth century have been fascinated by these deviations from strict perspective.

In the 1970s, TV engineers figured out several ways to warp images. Using analog equipment, they were able to warp an image to fit on a rotating sphere or a 3-D cube. An analog device can do any calculation you design it for, at close to the speed of light. These wonderful contraptions can warp thirty frames per second easily for real-time motion.

This effect is now commonplace, yet comparable digital effects have lagged. Computers powerful enough to handle that many computed pixels per second are only now becoming available.

So why don't we all have analog video devices? The reason is that every special effect requires another analog circuit. If you want a new effect, you need to build the hardware. With digital effects, you just need to write some code. The digital effect will never run as fast as the analog effect, but it is much easier to produce, debug, and modify software than hardware.

TELEVISION DISCOVERS MORPHING

Television, that great wasteland, really does have a purpose. It is to sell you things. And if you have to sit through an occasional boring melodrama, it's worth it to see the best of television—the commercials.

The commercials have pioneered many morphing effects. If you have only thirty or even fifteen seconds to sell your product, you go for the best impact you can get. Hang the expense. You're going to get killed by the air-time costs anyway, so get the best bang for your buck.

In the beginning were the traditional dissolves and optical effects, where kitchens would magically become spotless, and food would evaporate from the plates. Some of these effects are perfect illusions. At first they were startling, but viewers quickly gained sophistication. The effect was assimilated into the expanding vocabulary of visual phrases.

Then came analog effects. Torsos were twisted by stomach-aches and foreheads were warped by headaches. These effects are very realistic, and they cranked expectations up even higher.

Ultimately, computers started contributing to the visual lexicon. The first breakthrough was digital video. Once the image was in the digital realm, computers could work their magic—like warping and morphing.

The first warping for commercials was done by Pacific Data Images (PDI, Los Angeles). They wanted the Plymouth account, so they ran down to their local Plymouth dealer and picked up some brochures. They scanned different models into the computer and morphed between them. Plymouth was sold and PDI made a commercial that shows different parts of a Voyager warping as the voice-over comments on them.

Figures 3.2-3.4. *A few frames from a Pacific Data Images commercial for Exxon. Notice how the tiger's stripes are reflected in the wrinkling of the car. (Images courtesy of Exxon.)*

A later PDI spot also involved cars. In a commercial for Exxon, PDI made a car morph into a running tiger. The effect is made even more dynamic by warping the car to match the tiger's stripes (see Figures 3.2-3.4).

Kool-Aid has created a dazzling series of commercials where everyone and everything gets morphed to pieces. Schick commercials feature blockheads that metamorphose into normal faces (that need a shave, of course). Sega commercials have people's eyes bug out, and Nintendo warps kids flat with their G-Force commercial.

Michael Jackson's *Black or White* video put a spotlight on morphing. In a scene considered long by morphing standards, actors dancing in front of the camera are morphed, one after another. Faces of all nationalities are blended so well it's hard to tell where one begins and another leaves off.

Strangely, the rest of the video lapses into pointless violence. After its much ballyhooed premier, this part of the video was excised. With it went another morphing sequence between Michael Jackson and a black panther. It is competently done, but lacks the punch of the dance sequence.

THE MOVIES

Because of the expense of the method, morphing is used sparingly outside of commercials. Except, of course, for big-budget blockbuster movies. And who better to start the ball rolling than George Lucas?

WILLOW

In 1988, Ron Howard wanted some magical effects for his movie *Willow*. In one scene, the title character must metamorphose the good sorceress Raziel back into a woman. His first attempts don't work out so well, and he succeeds only in turning her into a crow, then into a goat.

Finally, putting everything into it, he tries again. Raziel's goat-neck stretches out to an impossible length, then she turns into an ostrich. Peacock feathers appear, then shrivel away. Willow keeps trying. Raziel shrinks down to a turtle, whose long neck pulls back into its shell.

She exhorts Willow to try harder, which he does. The turtle sprouts paws which grow into a tiger. Finally, the tiger metamorphoses into the exhausted sorceress, played by Patricia Hayes (see Figures 3.5-3.9). This amazing sequence was the first to expose an audience to computer morphing.

When director Howard told the special effects people what he wanted, they first toyed with the idea of cross-dissolving puppets, a time-honored tradition in the business. But Doug Smythe at Industrial Light and Magic (ILM, a division of LucasArts), remembered the short movie that Tom Brigham had made a few years earlier, and decided to give the computer a try.

He wrote a new version of the warping and morphing software. It was decided, after a few

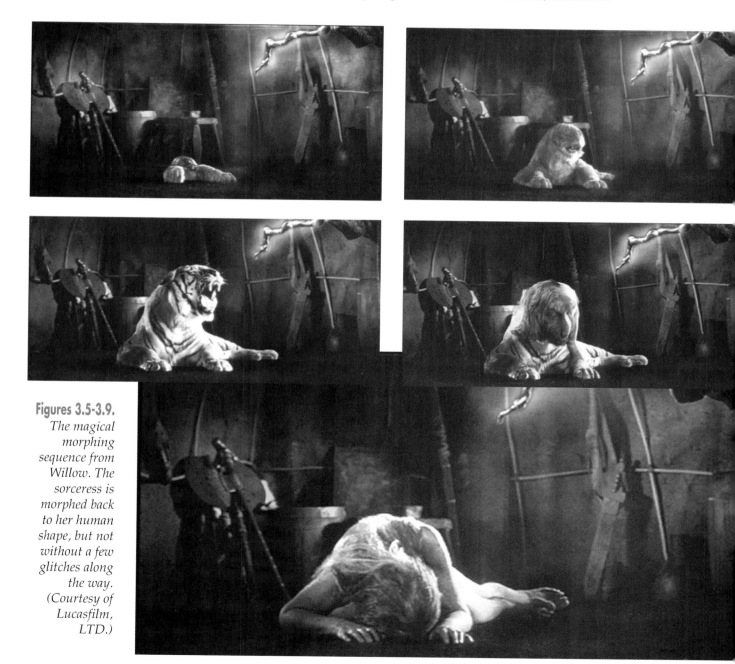

Figures 3.5-3.9. *The magical morphing sequence from Willow. The sorceress is morphed back to her human shape, but not without a few glitches along the way. (Courtesy of Lucasfilm, LTD.)*

trials, that the software couldn't do everything. Some of the stretches were so exaggerated that the image became pixillated—the pixels themselves were stretched to the limit. So a compromise was devised.

Models were created that could do part of the warping, then the computer would take over and finish the job.

For the goat, a clever puppet was made that could stretch its neck and do some of the metamorphosing on its own. That was accomplished by building a foam goat head around a fiberglass cast of an ostrich skull. When a vacuum inside the skull was turned on, it sucked the foam into the shape of an ostrich head. At the same time, the neck was elongated.

The models were filmed and then the programmers and animators went to work. They further stretched and warped the images, and finally pieced them together with the morphing algorithm.

The movie was a hit, and audiences begged for more morphing. They were soon to get it.

The audiences weren't the only ones to be impressed by this tour-de-force of special effects. ILM's technique garnered them a well-deserved Academy Award.

INDIANA JONES AND THE LAST CRUSADE

The next director to pick up the morphing baton was Steven Spielberg. He called on the now-experienced morphing team at ILM to craft the dramatic demise of the villain in *Indiana Jones and the Last Crusade*.

After chasing around the world to find the Holy Grail, Indiana Jones and his evil rival converge on the prize at the same time. Confronted with a choice of goblets, the scoundrel grabs what he thinks to be the Grail.

Drinking from the Grail is supposed to confer immortality, so he chugs down the contents of the goblet. Whoops, wrong goblet. The blackguard is sent to his just reward as the poison quickly ages him—then it kills him and dries him out for good.

The sequence involved lots of makeup and three progressively mummified latex masks. Then the computer folks massaged the images to smoothly morph each face into the next.

The result could make a person give up drinking entirely.

THE ABYSS

After some films that he probably would like to forget, in 1989 James Cameron made a movie called *The Abyss*. One of the key players was a computer-generated water "pseudopod." A lot of computer power was thrown into this movie.

The movie concerns the rescue of a Navy submarine from its precarious perch on the brink of a bottomless trough (of course). Fortunately for the Navy, there is a manned, underwater drilling rig nearby.

After the requisite string of disasters and strange happenings, the crew of the Deepcore drilling rig works toward the precipice. Along the way they confront the pseudopod, a beautiful, computer-generated tentacle of water that undulates through their underwater habitat.

The pseudopod seems to inspect one of the crew, played by actor Ed Harris. In a startling transformation, it takes on Harris's face. This terrific morphing effect is the highlight of the movie.

The computer work was done at ILM, of course. The graphics supervisor was Jay Riddle, working with Scott Anderson (no relation, but a nice guy), Lincoln Hu, Mark Dippé, Steve Williams, and John Knoll.

Using Silicon Graphics workstations, they spent six months designing the watery creature. They used a software animation system from Alias Research in Toronto, Canada. It was chosen because it uses spline- or curve-based shapes. Water just doesn't look right modeled with polygons.

ILM also used Renderman, from Pixar in San Rafael, to create the final images. The feature that attracted them to Renderman was its capacity for "motion-blur." This useful effect corrects a problem with all previous animation: jerkiness. When you animate, you create crisp, clean images. But real motion sometimes leaves a blur on even the fast exposures used in film.

Renderman mimics this effect, blurring the action when the motion is faster than a certain threshold. The effect also is used to match computer animation with the rest of the scene. In a quick pan, for instance, it won't look right if the computerized parts are crystal sharp while everything else is a blur.

The metamorphosing in this movie is three-dimensional, using carefully designed models. To make the pod imitate Ed Harris, the actor had to be scanned with a 3-D laser digitizer from Cyberware. Because the pod mimics his expressions, Harris had to be scanned while making different faces, as well.

The Cyberware scanning information can be used to directly create a 3-D object in the computer. It turns out, however, that it is very difficult to deal with realistic facial animation in 3-D. Instead, the data was "unwrapped" to create a 2-D-representation of the face. That image was manipulated by warping algorithms.

Each of Harris's expressions became a key-frame. Then warping was used to smoothly tween from one expression to the next. When the art directors were satisfied with the motion, the 2-D model was rendered as a 3-D surface for filming.

The folks at ILM had a great time on the project. It appeals to the Dr. Frankenstein in us to create impossible beings. Director Cameron was pleased, as well. The pod had made the leap from his imagination to the screen completely intact. That is one of the beauties of working with computer rendering systems. They can present anything, even dreams, with startling reality.

TERMINATOR 2: JUDGMENT DAY

If Cameron was happy with the effects in *The Abyss*, he was thrilled that they were brought in under budget and on time. It convinced him to give computers a bigger role in his movies. In 1991, he made the most technologically advanced movie to date. He wrote the script and directed the movie. He called it *Terminator 2: Judgment Day*.

Starring Arnold Schwarzenegger, the movie was a guaranteed box-office hit. Industrial Light and Magic again was picked to provide the computerized effects.

The story picks up more or less where the first *Terminator* left off. Schwarzenegger, as the Terminator, comes from the future again, but this time as a good guy. Again, Linda Hamilton is recruited to save the world.

She has her work cut out for her. The bad guys from the future have created a new, improved villain, the T-1000. He is a gleaming liquid-metal, quick-change artist. The role is sometimes played by actor Robert Patrick, but just as often it is the computer that you see on the screen.

For this magnificent antagonist, the wizards at ILM created a complete three-dimensional model of Patrick. They analyzed his gait and gestures. All of this went into the computer. What they cranked out seemed to be a living, breathing human, but it was as evanescent as the bits that made it up.

Finally, here was an actor that was not bound by the rules of the screen-actors guild. He could walk through fire and steel bars and not break a sweat. For director Cameron it was literally a dream come true.

Tweening, of course, gave this computer actor his smooth motions. Metamorphosing was used to produce the creature's changes—such as the one where he disguises himself as a linoleum floor, only to rise up and kill an unwary policeman.

Besides the 3-D creature, there is a lot of 2-D morphing happening in this movie. At one point in the story, the T-1000 impersonates Linda Hamilton. In the climactic steel-mill scene, he changes back to his normal shape as Robert Patrick.

The metamorphosis is perfectly executed, with Hamilton's hair pulling in then growing short as the morph unwinds. First one sleeve of the T-1000's uniform appears, then the next. All the while the features of her face slowly change into Patrick's.

One of the interesting aspects of computerized morphing is how transparent it seems. The effect is compelling even under the scrutiny of slow motion. In this video age, an effect like that can translate into big bucks.

It is one thing to morph Hamilton into Patrick. At least they had obvious points in common to map to. But there is another scene that has to be seen to be appreciated.

In a knock-down, drag-out fight with Schwarzenegger, Patrick gets thrown against a wall, face first. Instead of turning around to continue the brawl, the T-1000 simply *mutates* to face backward.

The job of the animator here was to make Patrick's face sprout out of the back of his head. Not having anything but hair to go on, they gave the scene a little punch by starting his features small, in the middle of his head. As the features grew to normal size, the hair faded out and the transformation was complete.

To keep the eye busy (and for those people with an itchy slo-mo finger) they even morphed his shirt. Two seams down his back smoothly shift in to the middle and become the seams around his front buttons. It is the only morph in the movie that isn't seamless (yes, that was a pun—I apologize).

The grand finale of the movie has our battered hero finally dispatching the T-1000 in a vat of molten steel. Steve Williams and Andrew Schmidt at ILM threw themselves, and everything else, into the task of animating the monster's swan song.

As he writhes and twists in burning agony, the T-1000 goes through all the forms it adopted in the movie. Heads and hands intertwine and fold into each other. In its final death spasm, the screaming creature turns inside-out and is gone.

Since the seminal *Willow*, morphing in the movies has grown exponentially.

In 1992, the Steven King movie *Sleepwalkers* (directed by Mick Garris) made good use of the morphing technique. The sleepwalkers of the title are "shape-shifters"—some sort of unholy cross between humans and cats. To live, they need human flesh—from pretty young virgins, of course.

Another recent movie to feature morphing is *Freddy's Dead: The Final Nightmare*. Here morphing is used to transport a house into outer space. With the house stationary, the backyard zooms away, morphing into the swiftly receding earth.

Finally, there were several morphing sequences in *Death Becomes Her*. Some pretty gruesome things were done to Meryl Streep and Goldie Hawn with morphing and warping. Streep's head gets twisted backward and Hawn's mid-section gets blasted out.

Of course they have taken a potion that makes them immortal, so they can suffer these indignities with something approaching aplomb. The potion provides another morphing situation as Streep grows younger in front of a mirror.

MORPHING ON YOUR PC

And now, morphing is not just for the big studios. Several companies are busily creating and releasing sophisticated morphing applications for the PC. These programs are surprisingly inexpensive, often costing less than two hundred dollars.

On the CompuServe network (1-800-848-8990), there is a shareware program called DMorph (for Dave's Morph) written by Dave Mason. This is a good place to start. With shareware, you can play with the software before committing to buy it. You can download it from section 13 of the Graphics Developer's Forum (GO GRAPHDEV).

Another software package is *Morph*, from Gryphon Software (1-800-795-0981). The original platform for this program is the Macintosh, but an IBM version is in the works.

Black Belt Software (1-800-852-6442) has a Windows version of morphing software called *WinImages:Morph*. Like the program from Gryphon, it uses one window for the source image and another for the target image. As points are entered for the source, they are replicated on the target image. By moving the target points, you define the warping.

Figures 3.10-3.13.
Four stills from a morph performed with the software included with Morphing Magic.

Photomorph, from North Coast Software in New Hampshire, is another Windows-based program. These PC-based software programs can deliver results rivaling the top studios, although not as quickly.

All of these programs use dot-mesh algorithms, unlike the vector-oriented approach pioneered by Pacific Data Images. To experiment with the vector approach, look at *Morphing Magic*, a book and software combination from Sams written by yours truly (see Figures 3.10-3.13).

With the price so low and the interest so high, expect to see a lot more morphing software in the near future.

Epilogue

Where are we today with morphing? Has it been done to death? Perhaps. It didn't help *Death Becomes Her*, which used morphing to great effect, but was critically panned and performed anemically in distribution.

Nevertheless, people have been in love with the idea of metamorphosing since the beginning of recorded history. As a literary or film device, it will always be popular.

The real legacy of morphing, however, is more subtle. It can be used to fix makeup or slightly modify some part of an actor. It can alter backgrounds and hide the wires supporting a stunt man.

It also can be used to cover up glitches. Say you have just shot two takes of a scene—you hate the beginning of one and you hate the ending of the other. The solution: take the good parts and splice them together with an invisible morph.

That's where morphing becomes an indispensable tool—it ensures the continuity needed to fool the eye. And that's what Hollywood is about.

Hollywood is also about money. It takes vast resources to make a movie. Unfortunately, that puts the power into the hands of the bean-counters. They want accountability, and in Hollywood, there are a lot of evanescent qualities—such as star power or location shooting—that are hard to pin down. Not so with special effects. There is a known amount of bang for your effects buck. This totally mercenary philosophy will ensure a steady stream of novel effects.

Morphing is here to stay, and now that it's here, watch out! Just remember, you can no longer believe what you see—at least in the movies.

Scott Anderson is the author of *Morphing Magic*, published by Sams.

CHAPTER 4

NANOTECHNOLOGY

Putting the ATOMS Where We Want Them

gayle pergamit & chris peterson

nanotechnology: the creation of microscopic-sized machines.

Imagine being able to cure cancer by drinking a medicine stirred into your favorite fruit juice. Imagine a supercomputer no bigger than a human cell. Imagine a four-person, surface-to-orbit spacecraft no larger and no more expensive than the family car. These are just a few of products expected from nanotechnology.

If you want to bet on what technology will be important in the future, experts agree that if you choose nanotechnology, you'll be a big winner. Compared to fractals, virtual reality, and other technologies making news today, nanotechnology is different in two ways. First, it is a technology just entering development: the first commercial products are still years away. Second, the scale and impact of nanotechnology will be immense. Rather than being an interesting single technique or application, nanotechnology will be the basis of humanity's next great technical expansion.

MIT's Marvin Minsky, well-known computer scientist and artificial intelligence pioneer, says "Nanotechnology could have more effect on our material existence than those last two great inventions in that domain—the replacement of sticks and stones by metals and cements and the harnessing of electricity."

Nanotechnology is the ultimate control of matter—the atoms and molecules that surround us—which we have worked to master since before our species earned the name *Homo sapiens*. Finally, the end of that long labor is in sight: in a short while, we will be able to manipulate the molecular structure of the objects around us as easily as we type a character on a computer screen.

Interest in nanoscale objects goes back decades to Erwin Schrödinger, who wrote about molecular machinery in his 1944 book *What is Life?* Later, in 1959, physicist Richard Feynman gave a talk titled "There's Plenty of Room at the Bottom," in which he pointed out that "The principles of physics, as far as I can see, do not speak against the possibility of maneuvering things atom by atom." Knowledge of nature's molecular machinery exploded in the 1960s and 1970s as molecular biology came into its own. The full picture of nanotechnology's promise became clear only in the late 1970s, when Eric Drexler, then at MIT, developed the concept in depth: nothing less than a complete manufacturing capability able to make objects to atomic specifications, with *each atom in a designed location*.

To convey the power of this technology, let's compare it to another technology regarded as revolutionary: the computer. Once the world had analog devices only, such as tape recorders using magnetic tape. Copies of the tapes can be made, of course, but there's a problem—as one makes copies of copies, the noise level gets worse and worse, until eventually noise is all that's left; the signal is gone. Analog systems tend to be unreliable: after a few steps, things go wrong.

Then came the digital electronics revolution, in which each bit is exactly right or exactly wrong, and errors can be made extremely rare. Suddenly, you could string together millions of steps with exact precision, creating objects of high information density: first the big clunky mainframe computers, then later computers just as powerful that fit on a desktop, or soon in a pocket. Compact discs are digital too—that's why their sound quality is better than the older analog LPs and tapes.

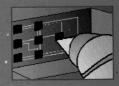

Why the "nano"? It's the next prefix in the pathway we've followed in the race for miniaturization: there was "milli," one-thousandth, then "micro," one-millionth. Now we're rapidly moving toward being able to do exactly what we want at the "nano" level (see Figure 4.1): one-billionth of a meter. Whoever picked this prefix got it right; "nano" comes from the Greek for *dwarf*.

Nanotechnology: Putting the Atoms Where We Want Them

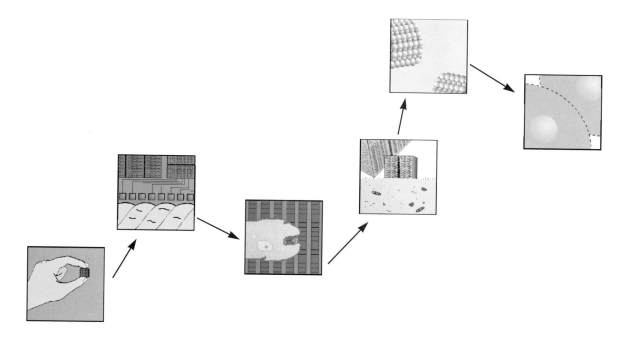

Figure 4.1.
Putting nanoscale in perspective.

Just as digital technologies handle information in basic units of a bit—which must be either a 1 or a 0, nothing in between—nanotechnology handles matter in its basic "bits": atoms and molecules. With the same kind of precision we see in the digital world, nanotechnology will let us put each atom in its designed place.

Just as computers changed forever the way we handle information, nanotechnology will change forever the way we handle matter, the way we *make* things. It's happened before: once *manufacturing* meant handmade; now it means machine-made, using machines on the meter scale. But machines already come in many sizes. Down at the molecular scale, molecular machines found in nature already build objects with atomic precision: the ribosome is "manufacturing central" for the cell, reading DNA and manufacturing complex proteins to spec. One machine produced by the ribosome is a bacterium's flageller motor, a full-performance rotary motor that powers the bacterium through liquid like a tiny speedboat. The aim of nanotechnology is to design and build our own new molecular machines, which themselves are able to build what we want, molecule-by-molecule. So watch the word manufacturing—someday not too far off it will mutate again, this time to mean *molecular*-machine made.

For most of us, it's hard to get excited about improvements in manufacturing—the subject conjures up images of smokestacks and clanking metal, thrilling in the 1800s, but today leading to worries about pollution or the tedium of the assembly line. But try a thought experiment: picture yourself in your house. Now subtract all of the objects except those that were made—or would be makeable economically—by hand. Unless you live like Davey Crockett, most of your things have just disappeared. You're left with wooden furniture and some clothes.

So a revolution in manufacturing can change our lives in ways hard to imagine in advance. But some basic points are easy to see already. With nanotechnology, the *range* of things that can be made will expand. The *quality* of those things will also increase, to the limits of our design power. And perhaps most importantly: the *waste* produced will almost disappear. With ultimate control, the uncontrolled materials we call "pollution" will become a thing of the past.

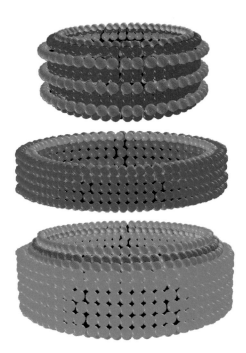

Figure 4.2. *An early design for a bearing with 2,808 atoms.*

What will a world of software engineers, chemists, medical researchers, mechanical engineers, and other designers create once they have this new set of capabilities? Some of the most popular choices include extremely small supercomputers, super-strong, lightweight diamond-like materials, and nanoscale medical robots capable of repairing cells to cure cancer and other diseases. The making of these products with extraordinary capabilities will be surprisingly ordinary: the practical side of nanotechmolecular manufacturing will be the old, familiar world of the factory, the hardware store, the workbench…but in nanoscale.

Welcome to the Nanotech Hardware Store

When built up from the basic "bits" of matter, gears can have meshing atomic-scale teeth. Bearings work with extraordinary smoothness, because they are designed such that the row of bumps of atoms on the working surfaces are precisely *unmeshed* as they come together (see Figure 4.2).

These gears, bearings, cams, rollers, belts, and struts will become the basic building components for molecular machines. If you put on your virtual reality goggles and dropped into the world of the nanoscale, you would find yourself surrounded by the familiar technology of conveyor belts and robot arms. It looks like any standard factory: machines stationed in neat rows, each conveyor belt bringing in its own sort of component, robot arms briskly grabbing a component and then popping it into place on the product under construction (see Figure 4.3). But here, the work goes on in a vacuum; the gear being slotted in place is a mere thousand atoms; the product is a supercomputer smaller than a human cell; and you're the only person on the shop floor.

To reach this world of molecular manufacturing, we'll need to develop a new kind of building technology: mechanosynthesis. Some people call it mechanical chemistry. It will be an extension of and possibly a replacement for much of the solution chemistry that is done today.

Today's solution chemistry—like yesterday's analog electronics—is a bulk technology, working with statistical populations of molecules. If you think back to chemistry class, you'll remember sloshing liquids around in test tubes. Working chemists slosh in clever, subtle ways, and often can slosh a product through a goodly number of steps before they run out of maneuvering room or errors accumulate and ultimately halt progress. When a synthetic sequence is ready for production, the sloshing is larger in scale and is done in vats. It's still bulk technology, whether done in a small tube or a large vat.

But with mechanosynthesis, we will have the ability to use a nanoscale robot to chemically "grab" the precise molecule desired and hold it where it is needed to form a new bond. For the first time, we'll be doing chemistry using "hands" instead of counting on diffusion to bring two reactant molecules together in the right orientation.

Nanotechnology: Putting the Atoms Where We Want Them

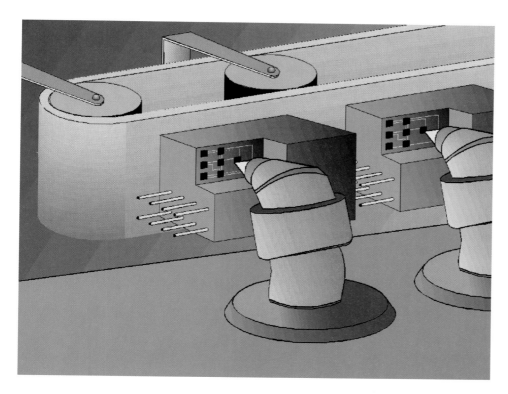

Figure 4.3.
Nanocomputers being put together on the nanoscale assembly line.

The big question today is "How do we get there?" There are two paths. In the first path, one uses mechanosynthesis to build molecular machinery, which could then be used to build better molecular machinery. To follow this development path, we would (1) develop the ability to perform mechanosynthesis, (2) design the molecular machinery (which could be done parallel with the first step), and then (3) build the machinery.

Path 1: Manipulating Atoms and Molecules

It took a few years for the invention of two physicists, Gerd Binnig and Heinrich Rohrer, to get the attention it deserved. In fact, one of the journals to which they submitted their paper announcing the new tool rejected it as "uninteresting," a decision they presumably regretted a few years later when Binnig and Rohrer won a Nobel Prize for the work: the scanning tunneling microscope (STM).

The word *microscope* is a bit misleading here because it implies that the tool is used only for looking at objects, not for changing them. But the tip of an STM is so sharp, and its positioning so exact, that in 1990 IBM researchers were able to use it to nudge 35 xenon atoms into letters spelling the company's name. The picture of "IBM" written in atoms showed up in newspapers around the world, alerting the public that the age of atomic manipulation was upon us.

mechanosynthesis: using a nanoscale robot to build molecular bonds.

Those atoms were just nudged along on a surface—no new chemical bonds were formed. But soon researchers started to use the STM to bond atoms in specified locations. The bare rudiments of mechanosynthesis are thus already demonstrated: mechanical and positional control of chemical reactions.

Using these proximal probes—the STMs and their newer cousins—it should be possible to do increasingly sophisticated manipulation and bonding. Special-purpose tips will be needed, such as a tip with an antibody to bind other molecules proposed by Eric Drexler, a research fellow at the Institute for Molecular Manufacturing (IMM) in Palo Alto, California. The path is clear: larger and larger molecular objects being built piece by piece, until designed objects such as molecular machines are routinely being produced.

The first part of this path—mechanosynthesis using proximal probes—is being pursued primarily in Japan, where the "Ultimate Manipulation of Atoms and Molecules" project is funded by MITI, the Ministry of International Trade and Industry. The government plans to spend $200 million over ten years, with an expectation of additional funds from the corporations participating. IBM is pursuing proximal probe technology also, with the goal of building extremely dense computer memories.

Path 2: Making Molecular Machines from Proteins

There's another path to nanotechnology and molecular manufacturing. Molecular machines in nature already perform synthesis of products, including making more molecular machinery. The cells of plants and animals are full of molecular machinery most of which, such as ribosomes, are already just as complex as the final products of molecular manufacturing.

Perhaps it would be faster to change these slightly, so that instead of producing more of the same old protein, they produce protein-like parts for molecular machines that we design. Then the challenge becomes designing molecular machine parts that will exploit the process of self-assembly to come together into working nanomachines. Such assembly is done in nature all the time—it's how the protein machines in our bodies are made—and should be possible to mimic, given enough time and design resources. If we design these parts cleverly enough, we could get them to self-assemble into machines able to perform mechanosynthesis.

So the two possible paths are: (1) mechanosynthesis first, leading to molecular machines, and (2) molecular machines first, leading to mechanosynthesis. Path 2 differs from Path 1 only in how the machine parts are made. Instead of building them with proximal probes, they would be built by the protein-producing apparatus of the cell, or by traditional organic chemistry.

It's hard to say today which of these paths will be fastest, so it makes sense to follow them in parallel. In both cases, the earliest results are not the sophisticated manufacturing systems, but cruder, simpler machines on the pathway. These machines could be quite useful as scientific instruments. For example, a system the size of a few proteins could read DNA, and a chip with many such machines could do the work of the decade-long, billion-dollar Human Genome project in a single afternoon.

The Importance of Design

Both paths are heavily dependent on the clever design of molecular machine parts. Two organizations in Palo Alto, California—the Institute for Molecular Manufacturing and Xerox Palo Alto Research Center—are now cooperating on the first design tools specifically aimed at molecular manufacturing. IMM researchers Eric Drexler and Markus Krummenacker are working on ways to design early building blocks. PARC researchers Ralph Merkle and Geoff Leach (visiting from Australia's Royal Melbourne Institute of Technology) have teamed with Drexler to produce a tool for the automatic generation of molecular machine structures.

The design task is being tackled from the other end as well—the design of parts for *advanced* molecular machine systems: gears, bearings, rollers, and other components not buildable today. These and other innovative designs are the first of what will ultimately be a huge nanotech hardware catalog from which designers will pick and choose.

Anticipating the Breakthroughs

Many people have the mistaken belief that we already have well-developed nanotechnology; this confusion is generated by the increasing number of products and technologies that flourish in the nanoscale regime but are not nanotechnology in the "atomically-precise" meaning of the term. But much of this current work in the nanoscale regime might provide benefits to molecular nanotechnology by increasing the tools and capabilities for working in that domain.

The announcement of any new "micromachine" provides another source of confusion: tiny motors and machines that sound nanotechish, but are a billion times larger in volume than a nanomachine. Micromachines are built the same way we carve, sculpt, or cast other things today—not by mechanosynthesis. Therefore, they are not the highly controlled and capable devices expected from nanotechnology.

People who wish to stay abreast of actual progress in nanotechnology and its enabling technologies, or participate in nanotech research directly, have an expanding set of resources. The 1992 publication of *Nanosystems: Molecular Machinery, Manufacturing, and Computation* by Eric Drexler gave the field its first text. Interdisciplinary study groups have been set up at schools such as MIT and Stanford University so that the chemists, physicists, biologists, computer scientists, and mechanical engineers needed for a full understanding of all aspects of the technology can share information.

The general public can learn about the impact of nanotechnology from an educational organization, the Foresight Institute, also based in Palo Alto, California. Potential uses, both positive and negative, are sketched out in the books *Engines of Creation* and *Unbounding the Future*. Interestingly, it is not nanotechnological accidents that most concern nanotech experts, but instead the controlled use of the technology for aggressive purposes. Like many powerful technologies before it, nanotechnology could be used to make improved weaponry, as well as improved consumer products.

THE BIG QUESTION: WHEN?

Once newcomers to the concept of nanotechnology have firmly grasped the concept and reviewed the state of current research, their first question is "When will all this be happening? Will development take years, decades, or more?"

There is no way to calculate when a technology will be developed. The rate of success depends on so many factors: how heavily the work is funded, whether it is done centrally or in a dispersed way, the quality of any interdisciplinary cooperation needed (which is a great deal in the case of nanotechnology), and the rate at which unexpected shortcuts appear.

That said, what is the best guess of those working in the field of molecular nanotechnology? The most frequently quoted number is about fifteen years, putting a major technological revolution before the year 2010. That year sounds far off, but a useful exercise is to calculate how old you or your children will be then. This change will happen within the lifetimes of you or those you care about.

Not One Technology, But Many

The concept of nanotechnology, the practicality of working at nanoscale, and the potential benefits of molecular manufacturing are beginning to capture increased research funds and the attention of more teams of computer scientists, chemists, biologists, physicists, and mechanical engineers. As researchers go about inventing nanoscale tools, designing components and machines intended for molecular manufacturing, and pursuing different development strategies in achieving nanoscale control, they will generate increasing diversity.

Figure 4.4. *An interdisciplinary team will be required to design and build this new medical tool, a device that removes foreign particles from the blood stream.*

Each new invention will bring with it opportunities for nanotechnology to diversify, specialize, recombine, and advance on unpredictable development and product pathways (see Figure 4.4).

As nanotechnology moves down these paths, it will acquire new names: surely computer scientists will wish to distinguish the molecular machines they use from the ones biologists use. In the far future, possibly none of it will actually be called "nanotechnology." It will simply be the technology that we use to make and do most of the things in the world around us.

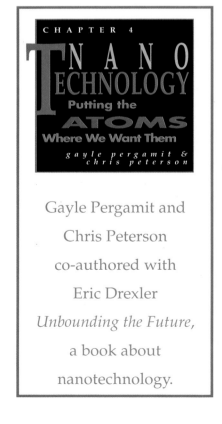

Gayle Pergamit and Chris Peterson co-authored with Eric Drexler *Unbounding the Future*, a book about nanotechnology.

CHAPTER 5
Smart materials
ivan amato

In 1959, William H. Armistead, then an executive at Corning Glass Works, posed a challenge to S. Donald Stookey, one of the company's most creative glass researchers. Develop a new kind of glass that darkens in bright light and that clears up again in dimmer settings, Armistead suggested. Within a few years, Stookey had come through with a material known as photochromic glass. By the 1980s, eyeglasses that could double as sunglasses were selling around the world by the millions.

> **smart materials:** materials that react to changes in their environment.

Stookey didn't know it, but his self-adjusting glass was one of the world's first commercial smart materials, developed more than a decade before scientists even began using the term smart materials. The smartness of the high-tech glass is a tango of light and chemistry. Sunlight entering the glass triggers a photochemical reaction that coerces silver ions mixed into the glass formula to temporarily extract an electron from nearby copper ions, another part of the formula. The electron exchange changes the formerly charged silver ions into electrically neutral silver atoms, which join into millions of tiny light-blocking clumps throughout the glass. When the light dims, the silver atoms in the clumps return their borrowed electrons to the copper ions and the clumps disband.

Stookey's invention might hold the honor of being one of the world's first money-making smart materials, but a growing community of researchers around the world are on the trail of a huge variety of sensitive and responsive materials, which they hope to build into what they call smart and intelligent structures. Their wish lists include items like these: self-healing concrete that can sense cracks in its bulk and seal them without human intervention; metal wires that change into different shapes, like flexing muscles, depending on the temperature; smart climbing ropes that change their colors (a kind of inanimate bruising) when strained beyond a safe amount; and military fabrics whose colors and patterns change like chameleon skin. Engineers are using materials like these to build experimental buildings, planes, and other structures that have decidedly lifelike qualities, beams Peter Gardiner of the University of Strathclyde in Glasgow, Scotland where he is director of the Smart Structures Research Institute (SSRI).

WHAT ARE SMART MATERIALS?

Just ten years ago, smart materials and structures were mostly futuristic ideas secretly probed in military laboratories. Not any more. Now they are the center of a bona fide research field with players in industrial, academic, and government labs around the world. The research even is starting to move beyond laboratories and into applications. Julian Vincent, a British zoologist at the Center for Biomimetics at the University of Reading in England says, "Intelligent behavior in materials and structures is fast becoming a reality."

Of course, when Vincent and his colleagues use words such as *smart* and *intelligent* in front of words such as *glass* and *concrete,* they do so with a subtle smile. These structures aren't smart quite the way living things are smart, although some researchers predict that the distinction will get increasingly blurry.

Look at a natural model—skin. Skin remains a perfect fit, no matter how big a person gets. It just manufactures more of itself as needed. During hard work, glands squirt sweat onto the skin's surface where it evaporates, taking with it excess body heat. Nerve cells in the skin sense heat, pressure, and the texture of objects. They also send pain signals to the central nervous system, which then activates muscles to get the skin away from whatever is causing the pain. Even when skin gets damaged despite these safety mechanisms, it repairs itself.

So in their most sophisticated embodiments, smart structures that mimic biological organisms would have ersatz nervous and sensory systems that sense what is happening within themselves and around them. They would have artificial brains programmed to interpret that sensory input and then to plan out responses that a human being might suggest.

To carry out those plans, these sensitive and perceptive constructions would need synthetic muscles and other active components that can do things—called actuators in engineering parlance.

Craig A. Rogers, director of the Center for Intelligent Material Systems and Structures at the Virginia Polytechnic University and State University in Blacksburg, concedes that neither he nor his colleagues in the smart structures business have made any artificial constructions as brilliant as skin or other biological tissues. But Rogers likes to remind people that biology's own smart materials have taken hundreds of millions of years to evolve in a natural trial-and-error process known as evolution. These are early days yet for smart materials researchers, concurs Vincent.

> actuator: an active component that can initialize responses.

To get structures closer to having a biological kind of smartness, they probably will require microprocessors to direct communication between and among their many sensors and actuators, says Edward Crawley of the Massachusetts Institute of Technology's Space Engineering Center. One favorite visionary example of his is a smart airplane wing that changes its contours, like a flexing arm muscle, to accommodate second-to-second changes in wind speed, air pressure, and other flight conditions, which would be measured by networks of on-wing sensors. The silicon brains would interpret the sensors' signals and orchestrate the actuators (see Figure 5.1). The continuous adjustments in wing shape, claim Crawley and other aeronautic engineers with smart materials on their minds, should up the plane's fuel efficiency while making the crafts safer. In a flight of fancy, Crawley muses that with smart structures technology, "you might one day have a plane that can land anywhere like a bird."

> How far along is the evolution of synthetic smart structures? Most examples so far involve structural components, such as reinforced concrete of buildings, that are rigged with sensors, such as optical fibers, that can serve like nerve fibers. Ditto for composite panels—layers of polymers laced with tough ceramic fibers or particles—of a satellite's solar panels or aircraft wings. In some instances, engineers are taking the additional step of including actuators into their designs to create structures that can sense and respond to the world by, for example, becoming stiffer in response to a load.

First things first though. And that means that Crawley and colleagues have tough challenges to meet, such as developing ways to embed microprocessors into composite materials without letting the high heats involved in manufacturing composites destroy the chips.

The challenges seem daunting, but momentum is growing. Until a few years ago, Crawley, Rogers, and their research kin mostly were working separately in different disciplines, including chemistry, physics, biology, and engineering, and were unknown to each other. But they now are congealing into the field of smart and intelligent materials and structures. In 1988, they convened in Japan for their first scientific conference. Since then, hundreds of scientists and engineers have met for subsequent conferences. The first technical journals devoted to the field emerged in 1989—another unmistakable sign that a new branch of science has sprung.

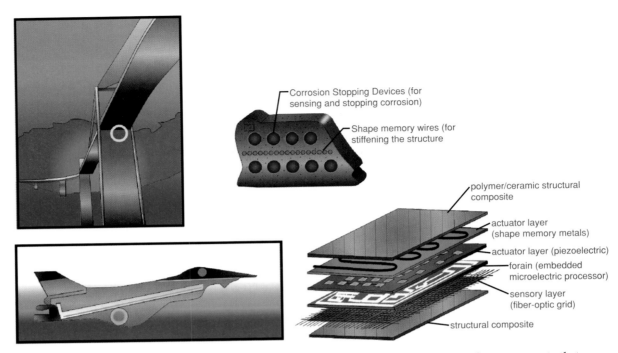

Figure 5.1. *A world of smart bridges, planes, and other structures will be made from complex components that harbor sensors, microprocessors, and actuators that can continually assess their environments and respond appropriately. The concrete of a smart bridge might, for example, sense corrosion and automatically neutralize it, and a smart airplane wing might sense aeronautical conditions and change its contours like a muscle to optimize its efficiency under the changing conditions.*

The United States was the first to set up research centers devoted entirely to smart materials and structures. VPI's Center for Intelligent Material Systems and Structures, founded in 1987, is the oldest and most well known. Since 1987 though, other places including the University of Michigan and the University of Vermont have opened their own centers. In addition, hundreds of scientists and engineers at the National Aeronautics and Space Administration (NASA), at dozens of high-tech companies, and at military laboratories where the idea of smart structures first emerged in the 1970s, are developing smart materials for everything from robots to nuclear waste containers.

Japan also has adopted an aggressive pace of smart materials research, having entered the fray in the late 1980s. Now, many Japanese researchers at universities and companies are working on projects including buildings that can respond to earthquake motions like commuters shifting their weight on subway rides and smart toilets that alert doctors if a user's urine contains molecules that might indicate illness.

Researchers in Europe, Australia, and countries throughout the world also have begun entering the field. In May 1992, the First European Conference on Smart Structures and Materials was held in Glasgow, where Peter Gardiner and his smart materials colleagues at the University of Strathclyde played hosts.

THE SMART MATERIALS VISION

Now that this new high-tech science with a sexy and exciting name like smart materials and structures is building momentum, an important question might come to the minds of those in government and industries that might fund the research. What's the goal? Surely civilization has survived without smart materials.

Craig Rogers offers one goal that Vincent and others second. The world of superhighways, massive skyscrapers, and high-speed planes harbor large-scale technology-based hazards unknown before modern times. In contrast to an ailing or damaged body, which can relay pain signals as a warning that something is amiss, artificial structures remain safe and healthy only if knowledgeable people inspect and maintain them on a regular basis. The smart materials crowd wants to begin designing buildings and vehicles and other constructions with an artificial brand of self-awareness. With smart, self-healing concrete, for example, buildings shaken during earthquakes might not collapse into piles of rubble, killing and maiming those inside. Or fatigue-sensing metal might prevent air disasters like the 1992 tragedy in which an El Al cargo aircraft crashed into a packed Amsterdam apartment building killing dozens of people because an engine mount failed.

If smart materials don't actually prevent disasters, they might ease the economic and social blow of unavoidable events. An infrastructure bolstered by smart materials might have weathered Hurricane Andrew of 1992 far better. The smart process here is akin to a people with sprained ankles shifting their weight to avoid further injuries.

To Raymond Measures, a researcher at the University of Toronto in Canada who works on self-monitoring airplane wings, the benefits of this kind of engineering are hard to overstate. He told the *New York Times* that it could become unethical in the future for engineers to design structures that are unable to sense impending failures. To do his part, Measures is developing sensors based on optical fibers that he hopes will render airplane wings smart enough to know when they are damaged and in need of attention.

Preventing disasters and shrinking their consequences are only two of the benefits that scientists expect from smart materials and structures. Others have more to do with the environment and economics. Smart-structure concepts could help engineers and construction firms use far less steel, concrete, and other raw materials in their constructions. That would make vehicles, skyscrapers, and other structures more energy efficient, possibly cheaper to build, and easier on the environment, Rogers says. Today, to insure that bridges, elevated walkways, and other structures are safe, engineers add tons of dumb but strong steel reinforcement. Equally safe structures could be designed by replacing a lot of that dumb brawn with smaller amounts of brainier materials rigged with artificial musculatures that can summon additional support when and where it is required, Craig suggests (see Figure 5.1).

MAKING VISIONS REAL

Sounds great, but realizing these dreams takes real sensors and actuators. It takes making real smart structures. Indeed, a growing catalog of sensors and actuators is beginning to tantalize more and more engineers, who have begun assembling them into the first, and perhaps historic, generation of smart structures.

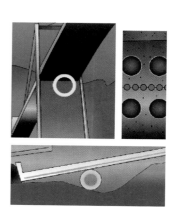

One of the most popular materials for a smart structure's nervous systems, the part that senses what is happening within or nearby the structure, are superthin optical fibers that conduct light much as metal wires conduct electricity. Mechanical engineer Tom Fuhr of the University of Vermont says that he and colleague Huston Dryver have riddled a bridge, building, and even a hydroelectric dam with optical-fiber networks that can track such happenings as vibrations and damage. Almost anything from changes in temperature to the onset of corrosion will change the properties of light—its brightness or wavelength, for example—as it travels in the optical fibers. Light detectors and computers that analyze these signals can reveal a detailed picture of the internal well-being of the structure, they say.

The dumbest way to use optical fibers as damage sensors is simply to note when they break because that will happen when an airplane's wing composite material breaks. Light in these broken fibers might even leak out to the wing's surface like blood and become visible to mechanics who would know where a repair was required. In a more sophisticated role, computers would analyze the sensory signals from the optical fibers and then send control commands to actuators that might strengthen the damaged wing section (see Figure 5.2).

Figure 5.2. *An advanced airplane with an optical–fiber-based health monitoring system embedded into its structure.*

> **piezoelectric: eliciting electrical pulses in response to mechanical perturbations such as changing pressures and vibrations.**

Another highly popular sensor material relies not on light but on electricity. Like the nerve cells in fingers, these piezoelectric (derived from the combination of the Greek word meaning "to press" and the word electric) materials elicit electrical pulses in response to mechanical perturbations such as changing pressures and vibrations. Polymers including polyvinylidene difluoride and ceramic substances such as lead zirconium titanate are piezoelectric. The molecular components of such material are arranged so that negative charges point one way and positive charges point in the opposite way. Pressure on the materials, therefore, brings the charges closer. That, in turn, increases the voltage between the material's opposite sides, resulting in an electrical signal. Researchers are looking into piezoelectric materials for earthquake-sensing buildings, stealthier submarine hulls that not only detect sonar signals but also quiet their echoes, and smart roads that can determine if a car or a big rig is driving on them. The government already is funding research for a gargantuan smart highways project that infrastructure specialists have been dreaming about for years, which presumably could solve many of the country's traffic problems.

Piezoelectric substances work as sensors by converting mechanical forces such as pressure into electric signals. That's only half of what they can do. They also work as actuators. When these materials are placed into an electrical field, say, by flanking them between electrodes, they transform electric stimuli into tiny motions, even sound-producing vibrations if the stimuli is cycled fast enough. In the smart submarine skin application, the piezoelectric materials sense sonar signals and immediately make the hull vibrate so that it can cancel most of the sonar signal. That would make for a far quieter submarine fleet than is possible today.

As actuators, piezoelectric materials are well suited for rapid, subtle motions such as vibrations. But shape-memory metals often are the choice when slower, but more pronounced, motion is needed. Below its so-called transition temperature, a wire of shape-memory metal (nickel-titanium alloy, or nitinol, for example) can be bent into some form. But when heated beyond the transition temperature, it will try to assume the shape that it had been in when it last was at that temperature, a metallic act of remembering. The shape change happens as atoms in the crystal grains of the alloy switch between different arrangements. If a straight length of that wire happened to be embedded into a panel or beam and then is heated (perhaps by sending electricity through it), the wire exerts a force within the composite as it tries to assume its higher temperature shape, which might have been bent in some way. The result: the stiffness or strength of a beam or slab can be controlled.

THE SMARTENING OF THE INFRASTRUCTURE

As their pantry of sensors and actuators has grown along with their skill and experience in assembling, smart-structures researchers have identified plenty of projects to pursue. Besides Stookey's photochromic glass, which stands out as an early smart materials arrival, some of the first smart structures now have been making their debuts, although mostly as prototypes.

Air Force engineers were among the pioneers of the smart-materials field, so smart aircraft applications are among the most developed. For example, a smart wing innervated with hair-thin optical fibers could warn pilots of impending structural problems if that light signal travelling within them is altered by cracks or corrosion or some other kind of materials stress or pain. Of course, in this version of the smart wing the damage is already done and a pilot might end up watching a warning light blink as the plane nose-dives earthward. Better still would be sensory systems that warn of subtle damage, which technicians could repair before a crisis. Indeed, Measures at the University of Toronto has been working toward that goal by etching the fibers before embedding them into an experimental plane's composite skin. That way, researchers can tune the optical fibers' damage sensitivity, giving technicians more of an idea of what might need to be done.

shape-memory metals: metals that when heated beyond a transition temperature, try to assume the shape they had been in when they were last at that temperature.

Moreover, light signals travelling in optical fibers can harbor signs of subtler mechanical and chemical changes in the plane's composite skin long before overt damage such as fractures and cracks occur. In that case, says Eric Udd, an engineer at McDonnell Douglas—one of the world's largest aircraft manufacturers—"you might have time to change your altitude (to relieve stresses on the plane) or decide not to take off." For years, Udd has talked up the bright future he envisions for aircraft fleets sensitized with nerves of glass.

In fact, his vision extends from the birth of composite materials to the end of their service lifetimes. Fiber optic networks embedded in these materials could serve as a cradle-to-grave self-monitoring system, he projects. And that could improve the safety and reliability of composite materials, which already are found in vehicles, spacecraft, and sports equipment, but appear slated for more widespread use in the future. During manufacturing steps, the fibers could help engineers monitor the composite's internal temperature and condition. After the composite component had cured, the same fibers would monitor the component's health as it was assembled into products. That's not all.

The very same optical fibers could then monitor the component's health status during the lifetime of those products. Finally, the fibers would signal the composite's end, when the material had aged or been damaged to the point that it could no longer reliably serve its structural role.

Aerospace engineers also are developing smart structures for spacecraft and space platforms. One of them, Benjamin Wada of the Jet Propulsion Laboratory in Pasadena, California, is working with colleagues on the components of what could develop into huge space truss works. The lengths of the structural members are adjustable via piezoelectric segments, yielding truss works that can be adapted for different purposes. By activating different patterns of piezoelectric segments, for example, mission control could change the truss works' geometry, say, from a space crane to a platform. Or, the segments, if properly controlled, could counteract the contractions and expansions that occur as the platforms circle in and out of sunlight during their orbits. That extra steadiness could clarify the vision of telescopes and enhance the performance of other orbiting instruments that require nearly absolute stillness.

In more down-to-earth directions, engineers also are taking steps toward smart buildings. As Fuhr of the University of Vermont sees it, entire cities of the not-so-distant future will become congregations of intelligent structures, each building, bridge, or road capable of monitoring its own structural health and taking measures to defend itself against the stresses and strains that the cities' bustle places upon them.

While the Vermont engineers are among the earliest to move smart structures from the idea stage into place within real concrete structures, Carolyn Dry of the Architecture Research Center at the University of Illinois in Urbana-Champaign is well on her way to raising concrete to even higher levels of intelligence. She and coworkers have been developing a concrete with additives that sense when cracks and fissures form and then repair the blemishes, hopefully before a catastrophic failure. Says Dry: "The main problems with standard concretes are that they are brittle, porous, and very dumb." One of her paths toward concrete smartness relies on polypropylene fibers filled with, say, methyl methacrylate, a strong and tough adhesive. A crack in the concrete also breaks the fibers from which the adhesive spills outward into the crack, halting the crack in its tracks before it can cause greater damage (see Figures 5.3-5.5).

Dry also is developing a version of smart concrete that battles corrosion instead of cracks. To pull this off, the fibers get filled with calcium nitrate, an anti-corrosion chemical. These fibers are deliberately made with pores and then sealed with a waxy coating, which dissolves when the nearby alkalinity increases, an indicator that conditions are ripe for the concrete's steel reinforcing bars to begin corroding. When the coating dissolves under these conditions, the calcium nitrate leaks out and corrosion is nipped in the bud.

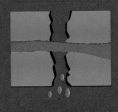

As a first step right on his own New England campus, Fuhr and colleagues have rigged a new building with optical fibers. Within its concrete and steel framework are several miles worth of fibers that will serve as a sparse artificial nervous system. One sensory job will be to detect and monitor vibrations, which could have consequences on the transportation and ergonomic studies that take place in the building's lower levels. How truck vibrations affect drivers' spines is one of studies on the agenda. The embedded optical fibers also might pick up disturbances from mild earthquakes, which sometimes rattle the area. They also will be able to track the health of the structure's concrete, hopefully helping the building's maintenance crew arrange for preventive measures and avoid unexpected, perhaps more expensive repairs. Fuhr and co-worker Dryver Huston also have innervated a nearby hydroelectric dam as well as a bridge along Interstate Highway 89 in Winooski, Vermont with optical fibers for measuring vibrations and stresses.

Smart Materials

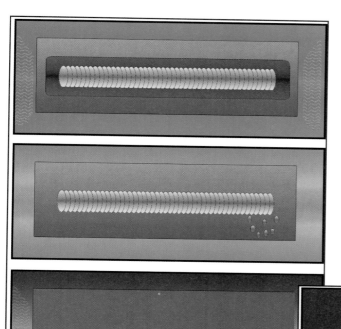

Figures 5.3-5.5.
Examples of smart concrete. There are two main tactics in the building of smart concrete. One method is to cover fibers with a waxy coating that dissolves under conditions that normally would cause steel reinforcing bars to corrode. When the coating melts, anticorrosion chemicals are released. The other method is to fill the fiber with an adhesive. When the fiber breaks, its contents spill into the crack and seal it before it can spread and further weaken the structure.

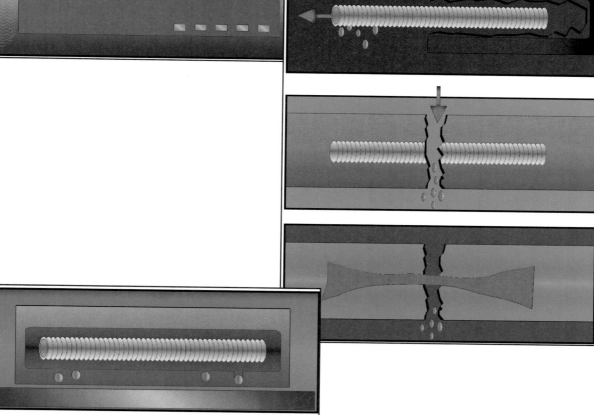

CHALLENGES TO HIGHER IQs

The field of smart materials and structures is so new that most of its challenges lie ahead. They won't be easy ones either. Researchers like Measures of the University of Toronto have shown that embedding optical fibers into composite components affects the composite's strength and performance in ways that are not yet predictable. Until smart materials advocates can convince aircraft manufacturers that smart composites will work as reliably and safely as conventional materials, the manufacturers will hesitate to use the advanced materials in place of more familiar and time-tested options. Similar requirements hold for embedded shape memory alloys and other actuating components.

Another challenge is to make the entire smart structure compact and light enough. At the moment, Crawley of MIT notes, computers external to experimental smart structures receive signals from sensors or send them to actuators. But aeronautic engineers won't be apt to consider smart airplane skins if that means they will have to include a roomful of computers to operate them, Crawley explains. That is why he and colleagues hope to develop ways of embedding electronic chips directly into the smart materials where they would be able to interpret the sensory signals and control the actuators without adding bulk and weight.

That scenario highlights additional challenges on a more fundamental level. Researchers still don't understand much about how signals from sensors, such as the fluctuations in light travelling through optical fibers or the electrical chatter from piezoelectric sensors, correspond to conditions of their host materials. Researchers need to understand those correspondences completely. Moreover, they also will have to formulate precise mathematical models of these relationships so that computers, or embedded microprocessors, will know how to analyze the signals. To be sure, smart materials researchers have a long way to go before they might flesh out the future they now envision. But if they do, the distinctions between the animate and the inanimate worlds, which now seems perfectly clear, may blur. With jets riddled with nerves of glass that fly like birds and bridges that feel pain and heal themselves, the humanly constructed world will harbor intelligent behavior previously reserved for the world of the living.

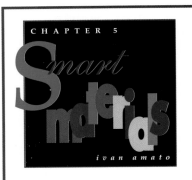

Ivan Amato is a staff writer at *Science Magazine.*

CHAPTER 6
3D anImatIon

mike morrison

Three dimensional computer graphics and animation have become mainstays for television and motion picture special effects. Although computer animation has been around for decades, its technology changes practically every month.

Some technological advances have come and stayed to become necessary tools for computer animation. Other advances have come and gone like fashion fads, and prove to be little more than mild curiosities. Although the future of computer animation is something very hard to predict, I will try to focus on those technologies that appear to have staying power.

This chapter delves into the cutting edge of 3-D computer graphics. In doing so, I focus on how computers simulate three-dimensional scenes and make them highly realistic by using advanced technologies such as ray tracing, metaballs, fractals, and much more.

Creating Artificial Realities

If the Great Masters of art who lived in centuries past had access to the technology available today, I have little doubt that some of them would have become 3-D animators. Why? It's a matter of the medium.

A painting is a fairly static medium. It is two dimensional, and it represents a fixed moment in time. Sculpture allows the artist to move into three dimensional space, but again it can represent only a fixed moment in time.

The field of 3-D computer graphics, on the other hand, allows the artist to create entire worlds within the computer. Not only can you create three dimensional models similar to sculpture, but you also can specify the color and position of lights and shadows, or even atmospheric effects such as rain and snow.

What really makes 3-D computer graphics special is that you can add motion. You can walk through the scene, fly the lights around, and the computer will automatically update with convincing realism the reflections and shadows of that artificial reality. These artificial worlds created within the computer's memory can be explored from any vantage point by the computer artist.

Using the Computer to Visualize Geometry

To create a 3-D image, the computer uses mathematical formulas to take a picture of a 3-D scene in its memory. A typical 3-D scene consists of the following elements:

- One or more 3-D Objects
- One or more light sources to illuminate the objects
- A Camera Viewpoint from which to view the objects

It is similar to photography. You need a subject to photograph (the 3-D objects). Next you need some light to illuminate the scene (the light sources). Finally, you look through your camera's viewfinder, line up the subjects, and click the shutter (setting the camera viewpoint and telling the computer to create the image). (See Figures 6.1 and 6.2.)

The computer artist has complete control over all of these elements. The artist can create 3-D objects in any shape, size, or color, specify the color and location of lights, and position a virtual camera. Based on the location of the camera, the computer will then generate an image that represents the 3-D scene as it would be seen from the location of the camera.

Once the 3-D scene looks correct to the artist, it is ready for movement with animation. The artist can specify a path for the camera to follow. Then the computer can be instructed to take a number of pictures as the camera moves along its path. Likewise, the artist can make any objects or lights travel along a path. When the finished images are played back from videotape or film, the result is computer animation. In this respect, the computer artist is part painter, part sculptor, part photographer, and part director.

3-D Animation

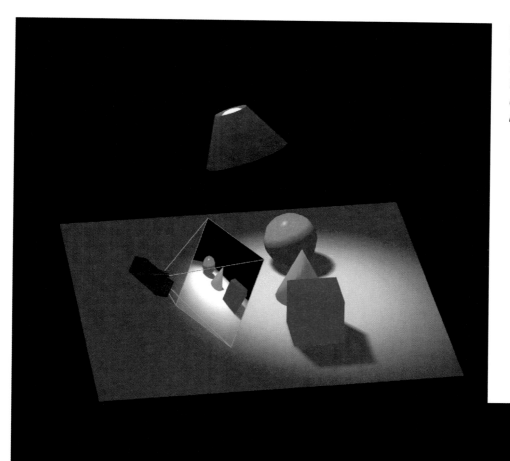

Figure 6.1. *The components of a typical 3-D scene. At the top a light source is depicted. On the left is a representation of the camera viewpoint. On the right are the objects.*

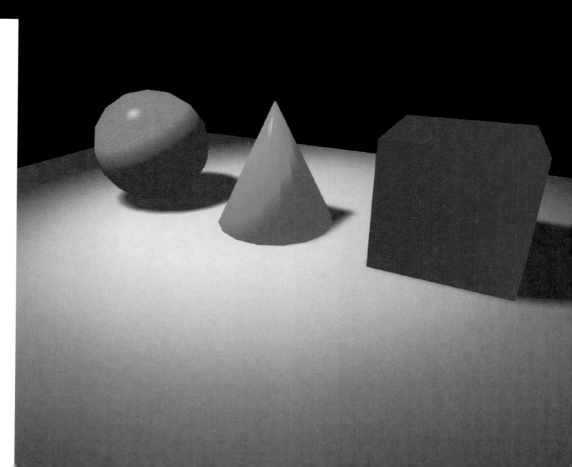

Figure 6.2. *The same scene when viewed from the camera's viewpoint.*

MODELING TECHNIQUES

3-D objects are created with a process called modeling. During modeling, you can create objects by instructing the computer to make simple geometric shapes such as spheres, cubes, cylinders, and so on (see Figure 6.3). These geometric primitives are fairly easy for the computer to model because they are defined easily with mathematical formulas.

Sometimes you can create complex shapes by combining these geometric primitives. At times, however, you need to create a complex 3-D object that requires advanced modeling techniques.

There are many ways to create more complex 3-D shapes. Perhaps the easiest way is to digitize the real thing. This is accomplished by shining a laser on the surface of the object while sensors detect the variations as the laser pans across the surface. This method is commonly used to create models of actors, where a computer generated version of the actor needs to appear for special effects, such as those seen in the movie *Terminator 2*.

> **modeling:** creating objects by instructing the computer to make simple geometric shapes, such as spheres, cubes, cylinders, and so on.

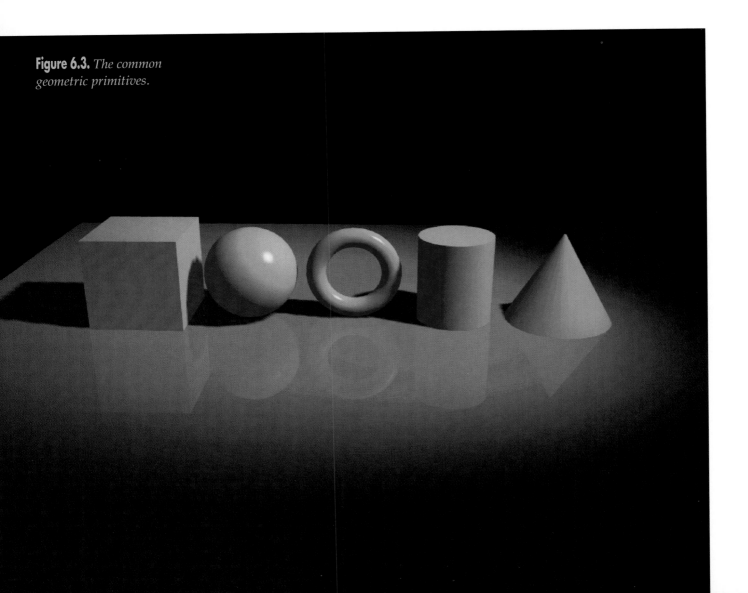

Figure 6.3. *The common geometric primitives.*

3-D Animation

Other ways of modeling complex surfaces involve fairly common techniques. Once such technique is called **Lofting** or **Extruding**. The term itself comes from ancient ship builders. To build complex hulls, they would first create cross sections of the hull. They built lofts to hold the hull during building and the process of hoisting the cross sections into the lofts became known as "lofting".

Say for instance that you wanted to create a 3-D cube by lofting it instead of using a geometric primitive. You would start by drawing a square on the computer screen. Then you would tell the computer to loft that flat 2-D square up into 3-D space. The computer would then stretch that flat square up into a 3-D cube.

Another modeling technique is the Lathe, sometimes called the Surface of Revolution. This is similar to the standard loft. However, instead of stretching the 2-D shape up in one direction, lathing revolves the shape around to itself. You can use this technique to create a wine glass quickly by simply drawing a cross section of the glass. Then you would use the lathe technique to revolve that cross section around 360 degrees (see Figure 6.4).

Some 3-D computer graphics programs allow you to draw a top view, front view, and side view of a 3-D object. The computer will then take these views and create a 3-D object for you.

A very popular way of creating natural, organic looking geometry is to let the computer generate it using Fractal algorithms. Fractal formulas enable the computer to generate random, natural-looking objects, such as mountain ranges, rocks, planets, plants, and trees.

A second common way of simulating natural phenomena is by using

> fractal algorithm: a formula that lets the computer generate random, natural-looking objects, such as mountain ranges, rocks, planets, plants, and trees.

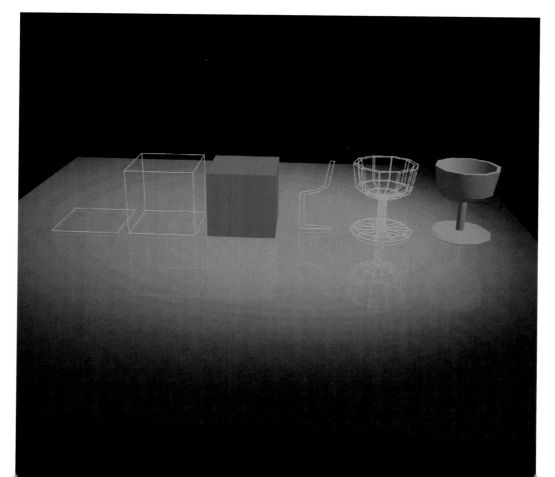

Figure 6.4. *Lofting a square up into a cube and using the lathe/surface of revolution method to create a wine glass from a cross section.*

Particle Systems. Particle Systems, are not real geometric shapes, they are collections of very tiny particles that move in any direction you specify. Particle systems can simulate natural phenomena such as splashing water, snow, sun flares, volcano eruptions, and so on.

A relatively new technology developed in 1982 allowed computer animators to create very lifelike, organic-looking models. The technology is called metaballs. This technique has just recently made it to the commercial fields of video and film. Metaballs are similar to drops of liquid that the artist can enlarge, stretch, and fuse together with other metaballs to form a naturally curving surface (see Figures 6.5 and 6.6).

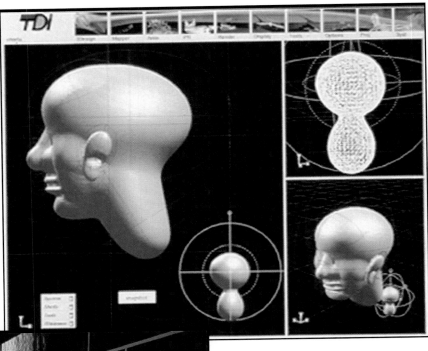

Figure 6.5. *Here is an example of a metaball modeling system.*

Figure 6.6. *A man modeled with metaballs. Notice the realistic, smoothly curving skin.*

After you have created your 3-D objects, the next step is to position them in 3-D space. The computer simulates a vast dark void you can populate with your 3-D models.

LIGHTS, CAMERAS!...

Now you are ready to set up some lights to illuminate your scene. There are four basic kinds of light you can use in a 3-D scene: Ambient, Omni, Spot, and Distant.

Ambient light simply defines the amount of background light in the room. Omni lights are similar to light bulbs. They cast light in all directions for a limited distance. Spot lights have source (where the light is located), target locations (where the light is pointing), and the width of the cone of light.

Spot lights are similar to invisible flashlights that you can position and point in any direction. Finally, distant lights point in a specific direction like spot lights, except that the rays of a distant light are all parallel, simulating the effect of a distant light such as the sun.

Finally, you have the camera. Like the spot light, it has a source (where the camera is located) and target location (where the camera is pointed). Most programs allow you to adjust the depth of field and the lens focal length for creating dramatic effects.

Finally, all the pieces of your 3-D scene are ready, as you saw in Figure 6.1. The final step is Rendering, where the computer creates an image that represents the scene from the camera's point of view.

RENDERING

Rendering can be the most time-consuming step in 3-D computer animation. It's where the computer has to perform mathematical somersaults to create a perspective representation of the 3-D scene you have created.

Rendering involves a number of complex steps. The computer must determine where the lights are, the shape of the objects, the properties of the camera (field of view, focal length, and so on), and many other details about the scene.

The computer then sets up a virtual screen and projects the 3-D scene onto that screen as seen in Figure 6.1. This final projection (also known as the rendering) is the end result.

HOW THE COMPUTER DOES IT

Let's discuss for a moment how the computer simulates solid 3-D objects. As I mentioned earlier, the computer artist has to create the 3-D objects themselves, usually in Wireframe mode.

Wireframes are much like the framework of a house or the steel work of a large office building. There are no solid walls, just a framework that specifies where the walls will be when the building is complete. Likewise the wireframe of a 3-D computer object specifies where the solid surfaces will appear when the object is rendered (see Figure 6.5).

Once the wireframe is built, the computer can render the object as if the wireframe were a solid object. The object's color will vary across its surface from light to dark based on the location and brightness of lights in the 3-D scene.

To further enhance 3-D objects, you can render them with Texture Maps. Texture maps are simply flat 2-D images that the computer wraps around the 3-D object. This is similar to applying wallpaper to a wall. To make a sphere look as if it's made of bricks, you would take a picture of bricks and apply it as a texture map to your 3-D object (see Figure 6.7).

Another method to enhance the realism is to use Bump Maps. Bump Maps take texture maps one step further by applying an image to simulate texture in a 3-D object (see Figure 6.7).

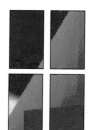

Figure 6.7. *Here is a rendered object in wireframe, solid color, texture mapped, bump mapped, and Ray Tracing.*

Finally, I come to reflective surfaces. There are three ways to simulate a reflective surface with a computer: Reflection Maps, Environment Maps, and Ray Tracing.

Reflection Maps take a single image and reflect it off the surface of the object. If the camera moves around the object, the reflection shifts realistically. For some applications this looks ok, but reflection maps do not reflect other objects in the scene and thus lose some realism.

Environment Maps are similar to reflection maps except the reflected image is created at the time the computer renders the object. The computer first renders a view from inside the object of the object's surroundings. Then it uses this rendering as the reflection for that object. The result can be very convincing because the object reflects other objects in the scene. There are limitations, however; if an object intersects the reflective object, it might not appear on the other object's reflection as it should.

To solve that problem you have to go one step further, into Ray Tracing. Ray Tracing simulates real-world optics by tracing every ray of light in a 3-D scene from the viewpoint of the camera through the scene to the ray of light's origin. This is the opposite of what happens with the human eye, where every ray of light starts from its origin and bounces off objects and into the human eye.

Because every ray of light is traced, the results of Ray Tracing can be spectacular. There are many added benefits to Ray Tracing, such as Refraction and Radiosity. Refraction is the effect of light shining through a solid object or a liquid. For example, looking through a fish bowl full of water causes the background to distort. Likewise Ray Tracing can simulate refraction effects.

Radiosity is a technique that simulates the way light bounces and reflects off other objects. Formulas similar to those that calculate heat dispersion are used to determine how much light a given object reflects. Radiosity can create beautiful images where scenes are illuminated with reflected light.

ANIMATING THE 3-D SCENE

Once the models, lights, and cameras are created, the last step is to animate the 3-D scene. As I mentioned in the beginning of this chapter, the artist can create paths for objects, lights, or cameras to follow. This is done by specifying Keyframes.

Say the animation is going to have 100 frames, and the artist wants object A to move to the right during those 100 frames. Instead of telling the computer exactly how much to move object A for each frame, the artist only needs to tell the computer where he wants object A to end up at frame 100. The computer will calculate exactly how much to move the object each frame, so that by frame 100, it is exactly where the artist wants it to be.

In this instance, frame 0 and frame 100 would be keyframes. Likewise the artist could tell the computer to rotate the object 360 degrees between frame 0 and 100. Then as the object moved to the right, it would also rotate as it traveled along its path.

In some cases, such as animating objects falling to the ground, the artist might not want to deal with moving the objects manually. In those cases computer artists can use Inverse Kinematics. Inverse Kinematics allows the user to simply specify that the object should fall. The artist needs to specify only how much gravity the computer should simulate, and the computer handles the rest of the animation by causing the object to pick up speed as it falls then bounce and tumble when it hits the ground.

While Inverse Kinematics offers a lot of power to the artist, it has its limitations in that you have little control over how the object bounces or tumbles. Inverse kinematics is controlled by the laws of physics, and sometimes those laws aren't aesthetically pleasing.

THE FUTURE OF 3-D GRAPHICS

As marvelous an artistic medium as 3-D graphics are, there are still a number of limitations. Currently, most 3-D computer graphics are created for two dimensional displays (such as the television or motion picture screen). Naturally, this leaves out much of the beauty of the full 3-D scene that the artist created. Also, the computer artist is usually forced to decide the best camera walkthrough for their scene, and the audience must follow suit.

However, with Virtual Reality looming on the horizon, these "lost masterpieces" might someday be rediscovered by virtual-reality explorers looking for new terrain. They will be able to fully appreciate the depth, motion, and atmosphere of 3-D animation being created today.

Mike Morrison is the author of *The Magic of Image Processing.*

CHAPTER 7
Welcome to the Virtual World

linda jacobson

Virtual reality: Why is it so *seductive*? Maybe because it's about creation. Or maybe because it represents our mastery of the world around us, through a process of techno-evolution that could help us solve major problems. Perhaps VR attracts our attention simply because it promises to bring satisfaction, joy, and new meaning to life.

One sure reason for VR's allure lies in its potential to enhance every field of human endeavor. Doctors use it to practice surgery without risking lives. Architects use it to test building designs before breaking ground. Artists use it to create new tools and new types of art. One day, kids will use it to meet historical figures and make new friends in other lands. Parents will use it to visit exotic locales without leaving home.

But is virtual reality as marvelous as the mass media and Hollywood would have us believe?

> **virtual reality:** technology that allows users to enter, manipulate, and travel through computer-generated, interactive, three-dimensional virtual worlds.

Before we can answer, we need to understand what virtual reality is, where it comes from, and where it's heading.

The technology of virtual reality lets you enter, manipulate, and travel through computer-generated, interactive, three-dimensional environments: "virtual worlds." Computer-generated means these worlds exist as models in the form of alphanumeric data stored inside a computer. The computer makes continuous calculations to display the worlds in visual form. Interactive means the virtual worlds respond to your actions. You use special devices to control your position and viewpoint within the virtual world. Three-dimensional means the virtual world and things in it seem to have depth as well as height and width.

Exploring a virtual world usually involves donning such trend-setting fashions as data gloves, head-mounted displays ("goggles"), stereoscopic glasses, or data bodysuits. To immerse yourself in a virtual world, you typically wear a data glove and goggles. They're wired to a computer system of software and hardware that calculates and displays a virtual world. The goggles contain two tiny video monitors that show a 3-D view of the virtual world—a space station, perhaps, or an octopus's garden (see Figures 7.1 and 7.2). The glove is lined with fiber optic wires, which detect the bend and flex of each finger (see Figure 7.3). Tracking sensors built into the glove and goggles send data about your head and hand positions to the computer, which responds by displaying the appropriate image.

As a result, you swing your head and see different objects and scenes above, below, beside, and behind you. Point your gloved finger and fly through the air, or clench your fist around a virtual object to grab or lift it. The goggles include headphones as well, so you can hear sounds—usually in 3-D. In some virtual-reality applications, we can interact with other people who are visiting the same virtual world.

The basic purpose of virtual reality ("VR") is to create a new type of human-computer interface. We use many types of interfaces every day, although we might not realize it. A door knob is the interface to a door. The interface to a car includes steering wheel, brake, gas pedal, and ignition. The Apple Macintosh revolutionized computing with its "graphical user interface"—it displays pictures of file folders and floppy disks that let you interact with the computer visually, rather than by typing commands. VR replaces traditional computer interface devices—keyboard and mouse, for example—with devices that allow more intuitive interactions with data and involve our senses in new ways. When we involve our senses in a learning experience, we understand more readily and retain more of what we learned.

On another level, VR is a communications medium characterized by interaction and a sense of "presence." One day we'll use VR to converse with, see, and touch people physically located many miles away.

Welcome to the Virtual World

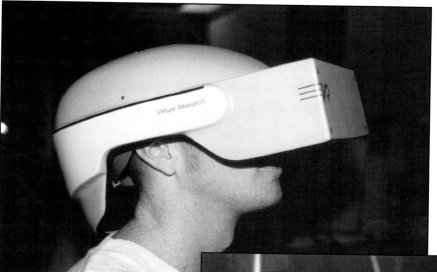

Figure 7.1. *Side view of head-mounted display by Virtual Research of Santa Clara, California. (PHOTO CREDIT: © 1993 Linda Jacobson.)*

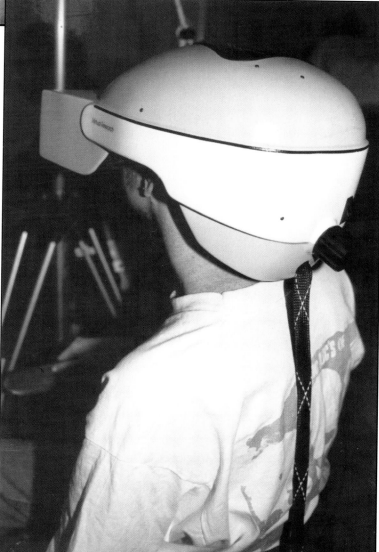

Figure 7.2. *Back view of head-mounted display by Virtual Research of Santa Clara, California. (PHOTO CREDIT: © 1993 Linda Jacobson.)*

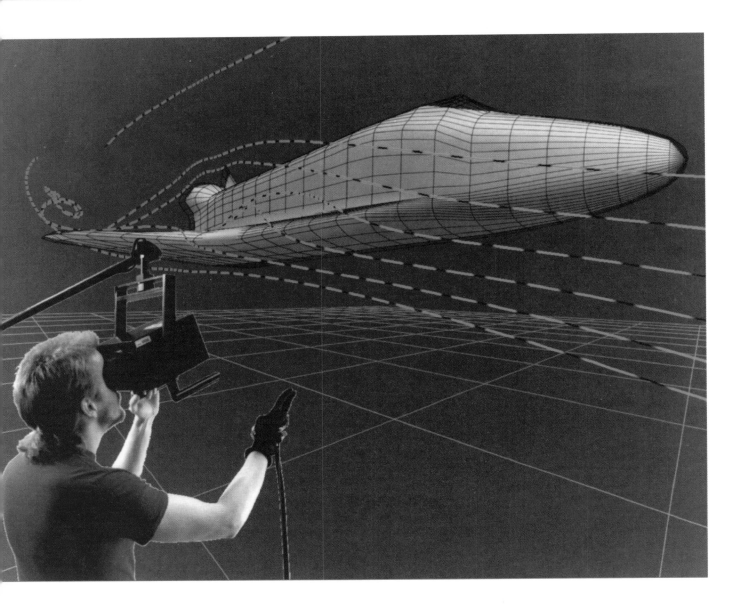

Figure 7.3. *Use of data glove and the "BOOM" stereoscopic display by Fakespace Labs of Menlo Park, California, to simplify access to the virtual world; in this case, a visualization of airflow dynamics of an aircraft wing.*

VR also is a tool for problem-solving. VR technologies "amplify" our intelligence by helping us turn data into knowledge. Like the telescope and microscope, VR provides a different way to see and experience information. It's one thing to scrutinize an X-ray to determine the best way to operate on a tumor. It's another to practice operating on a virtual version of the real tumor, understanding the effects before trying it with a live patient.

VR helps us solve problems because it helps us visualize information—it lets us view things spatially. When we can look at data in space as a whole or in detail, we see its elements in association with each other or individually. It's easier to solve a problem when it's presented spatially instead of on paper or computer screen. As a result, we make judgments faster than if we viewed the data in numeric or written form. It's much easier to pack a suitcase for vacation than to try to list everything you'll need and imagine how they'll fit in the suitcase.

Ultimately, VR is an artist's tool and medium. In virtual worlds we'll enjoy new ways to express and view ourselves and the world around us. Like art, VR supports different modes of perception and perspective, emotion and thought. VR lets us express concepts that cannot be said in words or in pictures alone. VR lets us become part of the art, co-creators in the artist's endeavor. "V-art" will be fun!

No matter how we define VR, the goal of virtual world builders is to create an experience that occurs in physical reality or only in the imagination, representing ideas so we can move through them and manipulate them in ways we cannot in physical reality.

One primary feature of a virtual world is immersion in the environment provided by a computer, the feeling of being inside that space. The experience involves the same mental shift of suspending your disbelief as when you get absorbed in a novel, film, or video game. The second main feature is interactivity with objects in that space; you can move things, open doors, walk around corners. The third main feature is the ability to navigate through that space, to go in any direction you want.

The term *virtual reality*, popularized in the late 1980s, probably won't remain trendy for very long. As more applications in business, educational, and consumer fields incorporate virtual world systems, technologies that now fall under the umbrella term "VR" will be absorbed into the mainstream of computer-based development and engineering. We'll probably identify the effects of the applications instead: synthetic digital environment, simulated buildings, 3-D terrain, immersion, and so on.

One VR-related term that probably *will* survive is "cyberspace." Cyberspace describes the "place" created by the networking of multiple VR systems and environments. We also use "cyberspace" to describe the information space in which we all operate. For example, some say that cyberspace is where you are when you're having a phone conversation. Cyberspace was coined in 1984 by science fiction author William Gibson in his book, *Neuromancer*. Gibson imagined that all who live by computers will one day commingle in a jointly created virtual world: "mankind's unthinkably complex consensual hallucination, the matrix, cyberspace…"

Virtual reality might seem like science fiction, but it represents the convergence of several nonfiction science disciplines, including human-computer interface design, simulation and data visualization, robotics, computer graphics, stereoscopy, and computer-aided design.

> **cyberspace: the area created by networking multiple VR systems and environments.**

Technically speaking, VR's primary ancestor is the simulation industry, which dates back to the flight simulators the U.S. Air Force started building after World War II. Some 15 years later, a young cinematographer named Mort Heilig saw Hollywood's giant-screen "Cinerama" films. The experience led him to develop and patent "Sensorama" in 1962, an early VR-type arcade attraction that combined 3-D movies, stereo sound, vibrations, aromas, and wind. The "passenger" sat on a motorcycle seat, grabbed the handlebars and peered into Viewmaster-type goggles to enjoy "the ultimate film experience." Sensorama was based in part on technology that Heilig patented in 1960, a viewing mask called "Stereoscopic Television Apparatus for Individual Use."

A few years later, a young engineer named Ivan Sutherland demonstrated to the scientific community the radical notion of using computers for design work. This paved the way for computer graphics, opening the door for hundreds of computer uses. In 1965, Sutherland wrote, "A display connected to a digital computer gives us a chance to gain familiarity with concepts not realizable in the physical world. It is a looking glass into a mathematical wonderland." This perspective led to his development of the first fully functional, head-mounted display for computer graphics—the helmet that led to today's stereoscopic data goggles.

stereoscopy: viewing an object so that it appears three-dimensional.

Stereoscopy—the viewing of objects so as to appear three-dimensional—dates back to 1832, when the stereoscopic viewer was invented. By the end of the 19th century, stereo viewers were all the rage. Today 3-D graphics is the field of computer science concerned with representing objects in three dimensions on a two-dimensional screen. Artists use computers to create complex 3-D images that can be changed almost instantly to suggest how your view would change as you move around, through, over, or under a scene. The 3-D graphics software imbues the image with visual cues that trick us into thinking we're seeing depth, or three dimensions. The image is flat, 2-D. When viewed in true 3-D with stereoscopic devices, the image seems to extend in front of and behind the face of the monitor. Imagine how a product designer can use this technique to visualize and examine every surface and line of an object before creating the real thing.

In virtual reality, 3-D imaging imbues the virtual objects with substance, and thus imparts a greater sense of presence (and helps suspend disbelief) in the virtual world.

Virtual worlds needn't appear three-dimensional to provide an enriching experience. An arts scholar named Myron Krueger wasn't concerned with stereoscopy when he coined the term *artificial reality* back in 1970. He was interested in "computer-controlled responsive environments" that approached computers from an aesthetic standpoint. Krueger felt that computer keyboards prevented most people from using computers for artistic expression. So he created "Videoplace," an art installation consisting of a computer-controlled video camera and large projection screen. As you face the screen, the camera below the screen captures your image, which is combined with computer-generated graphics and projected onto the screen. Your movements are translated into actions in the graphic scene. Your image, displayed in colorful silhouette, might lift, push, or toss graphic objects, swing on graphic ropes, and shrink, rotate, or move anywhere on the screen (see Figure 7.4).

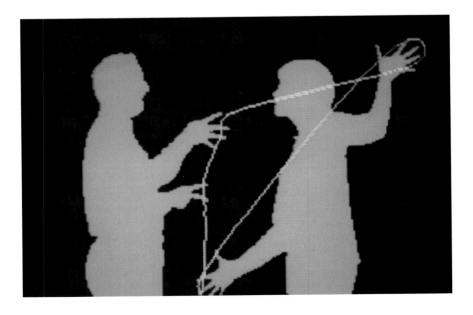

Figure 7.4. *Artificial Reality (also called Projection VR): This is "Cat's Cradle," an art-piece from Myron Krueger's Videoplace system. Here, graphic "string" is magically attracted to the finger-tips of the participants. (PHOTO CREDIT: Myron Krueger.)*

Throughout the 1970s Krueger struggled to find funding for Videoplace. Meanwhile, the federal government was backing other R&D projects that played roles in VR's history, such as MIT's Moviemap project (which let you "travel" through a videotaped version of Aspen by touching parts of the screen). Then there was the U.S. Air Force's SuperCockpit project. Marshalled by Tom Furness at Ohio's Wright-Patterson Air Force base, SuperCockpit came to fruition in 1981 after many years of development. The mock cockpit uses computers and head-mounted displays to depict 3-D graphic space through which pilots learn to fly and fight without taking off into real skies to inflict real injuries. SuperCockpit was amazingly successful. It cost many millions, however, and researchers at NASA Ames Research Center in Moffett Field, California, decided to develop a more affordable system.

NASA's resulting Virtual Interface Environment Workstation—developed mainly for planning space missions—was the first to combine computer graphics, video imaging, 3-D sound, voice recognition and synthesis, head-mounted display (based on video monitors taken from miniature TVs purchased at Radio Shack), and a data glove (based on an invention designed to play "air guitar" by two music-loving geniuses, Jaron Lanier and Tom Zimmerman). That's the data glove that caught the public's eye when it appeared on the cover of *Scientific American* in October 1987. The realization that NASA's achievements were based on commercially available equipment, and that devices like data goggles could be built with stock electronic parts, triggered research programs throughout the world.

Lanier and Zimmerman went on to co-found VPL Research in Redwood City, California, the first VR system supplier. VPL Research proved there was a market for virtual reality. Other companies (ranging from software firms to large computer corporations) soon started developing and selling VR. Today, some 65 companies in the United States alone are nourishing the burgeoning field. Some sell expensive products to the government, research institutions, and corporations; others develop affordable tools and systems for home computer owners. Over the next few years dozens more will join the act.

So what is the quality of a modern-day virtual world? That depends on the type of VR and accompanying input (control) and output (display) devices; the power of the computer supporting the system; and the purpose for the virtual world's existence. A VR game such as "Dactyl Nightmare" (which you can play on the goggle-equipped Virtuality arcade game systems around the country) looks like a primitive cartoon (see Figure 7.5), whereas a VR simulation of an air battle used to train military pilots seems frighteningly genuine.

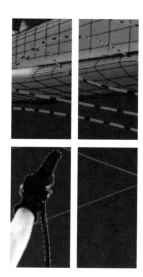

To explore a virtual world, you needn't be completely immersed in it, as Myron Krueger proved with Videoplace. In other words, you don't always need to wear goggles that cut off your vision and hearing from the real world. While there are many ways to define virtual reality, almost everyone agrees that in addition to "immersive VR," the technology comes in three flavors:

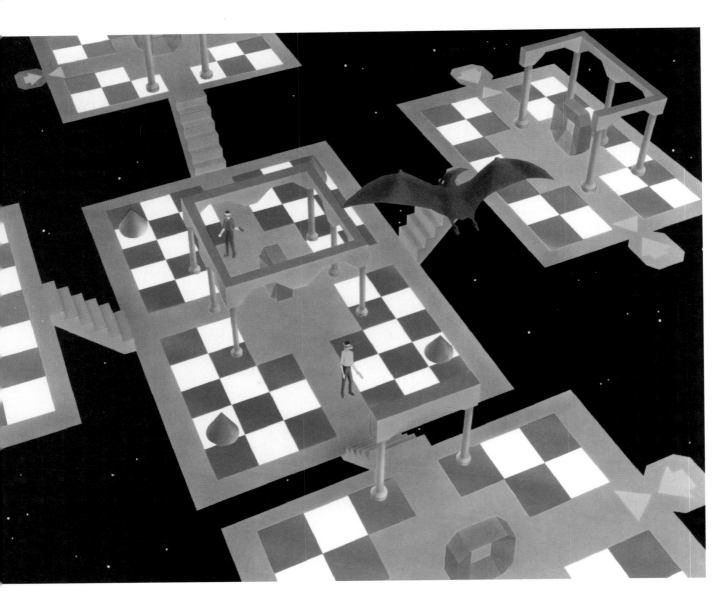

Figure 7.5. *The virtual world of "Dactyl Nightmare," the Virtuality arcade game from Edison Brothers Entertainment.*

Desktop VR (also called "homebrew VR" or "garage VR") is a low-cost method of interacting with a virtual world using a standard personal computer. The monitor serves as a "window" into the virtual world. Sometimes you view the world with wireless stereoscopic glasses that give the illusion of depth. You move your position and viewpoint by manipulating a joystick, keyboard, standard mouse, or 3-D mouse (see Figures 7.6 and 7.7).

Introduced in 1991 and discontinued shortly thereafter, Mattel's PowerGlove for Nintendo game systems made a big splash with computer programmers who connected the glove to their PCs to control graphic scenes. The result: a VR system (with choppy, slow graphics) that cost less than $2,000—a far cry from the $20,000 and more needed to configure "professional-level" VR systems.

Figure 7.6. *Desktop VR: A 3-D data visualization application using "CrystalEyes" stereoscopic glasses by StereoGraphics Corp. of San Rafael, California.*

Projection VR ("artificial reality") resembles Desktop VR, only the window into the virtual world is much larger and provides a better view for many people or a wider view for one person. You see an electronically generated image that you control by moving around inside the control space. Sometimes your own image is projected into the scene. You stand *outside* the virtual world but communicate with characters or objects *inside* it. Projection VR serves as a promotional tool at conferences and exhibitions, an educational tool in museums, and as a source of fun and folly in the entertainment field.

Simulation VR represents the oldest type of VR system, the kind in military tank and flight simulators and now in public game centers, such as BattleTech Center in Chicago. Simulation VR places you inside a physical mock-up of a vehicle that you control. Inside this cabin ("cab" or "pod"), video screens or computer monitors provide windows to the virtual world. You're not encumbered by cables extending from goggles or gloves and can interact with realistic physical controls. The cabs can be networked easily, allowing several people to participate in the same virtual world.

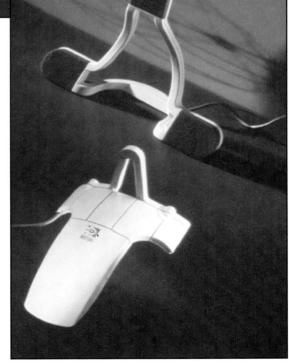

Figure 7.7. *The 3-D mouse by Logitech of Fremont, California. The mouse operates normally and also in 3-D mode (controlling objects on x, y, and z axes as well as in pitch, roll, and yaw orientations) for interaction with a virtual world. The device in the background tracks the mouse's position and orientation as you move it through the air.*

All VR systems rely on hardware but are useless without software that enables visual and auditory presentation, determining field of view, display clarity and speed, lighting, visual feedback, and physical dynamics of virtual objects (gravity, elasticity, weight, and so on). Research labs at corporations and universities currently are investigating the incorporation of artificial intelligence and artificial life-forms within virtual worlds.

Indeed, research labs devote long hours to exploring and developing virtual-world technologies and perceptual issues. Premier VR research sites include NASA Ames Research Center and SRI International (in Northern California's Silicon Valley), Human Interface Technology Lab at University of Washington/Seattle, Electronic Visualization Lab at University of Illinois/Chicago, University of North Carolina/Chapel Hill, University of Texas/Austin, Carnegie Mellon University, University of Central Florida/Orlando, and Banff Centre for the Arts in Alberta, Canada.

Meanwhile, professionals in disparate fields are climbing into virtual worlds. The military gets virtual for flight and combat simulation in training sessions. Stockbrokers enjoy applications for financial data visualization. In other areas, VR experimentation includes applications in education, medical imaging, molecular engineering, genetic engineering, medical diagnostics, and air-traffic control.

One early commercial VR package stems from computer-aided design for architects, product designers, and engineers. These professionals use VR to test the designs of everything from aircraft to cars to bridges to furniture. By creating virtual models of products, they can "fly" around them and get a feel for them. As a result, VR increases design productivity and quality, reduces time to bring a design to market, and cuts overall development costs (see Figure 7.8).

Figure 7.8. *A virtual world created with Cyberspace Developer's Kit from Autodesk of Sausalito, California. A simple scene such as this might be used by city planners to visualize a skyline for predicting the effects of a new building before construction begins.*

Sales and marketing executives love the potential of engaging customers with virtual versions of their wares. At Matsushita's Virtual Kitchen in Tokyo, shoppers select cabinets and appliances, then put on data goggles and gloves to "walk around" their new kitchen, moving the cabinets and appliances by grabbing and placing them where desired. When the customer is satisfied, a drawing program produces building plans, and the goods are delivered directly to the shopper's home.

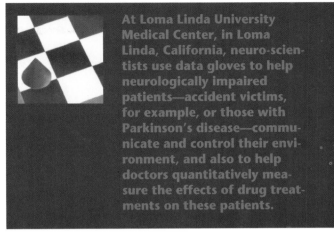

At Loma Linda University Medical Center, in Loma Linda, California, neuro-scientists use data gloves to help neurologically impaired patients—accident victims, for example, or those with Parkinson's disease—communicate and control their environment, and also to help doctors quantitatively measure the effects of drug treatments on these patients.

VR also holds great promise in the field of medicine. Stanford Medical School, in conjunction with NASA, created software for operating on simulated patients. They developed virtual bodies with physical calculations that provide accurate representations of body parts. Using gloves and goggles, doctors and med students experiment with procedures on simulated patients rather than real ones or cadavers.

The entertainment world is getting virtual, too. People are devising ways to harness VR to affect and enhance the illusory domains of cinema and stage, and to shape the future of the video game. Early VR entertainment ventures are enjoying great success in shopping malls, where kids and young adults flock to experience immersive and simulation VR systems, such as Virtual World Centers (launched as "BattleTech" in 1990) and the begoggled Virtuality game arcade system, which came to the U.S. from England in 1991 (see Figure 7.9). In 1993, Paramount Communications, Edison Brothers, and Spectrum Holobyte announced joint plans to launch "Starbase," a series of Virtuality-based entertainment centers featuring Star Trek's storylines and crew. Beam us up!

Figure 7.9. *Virtuality Center and players in Exhilirama in Crestwood, Missouri*

Meanwhile, theme parks around the world are gearing up to bring virtual realities to life, while motion picture companies are developing virtual movies, or "voomies," in which the audience interacts with the movie actors. Virtual popcorn, anyone?

By now you might wonder what has to happen before we all can soar happily into virtual skies. The challenges are many. One concerns the high cost of virtual reality. Immersive VR systems start at around $20,000 and can reach over a quarter of a million; however, do-it-yourself types can explore low-end "garage VR" for as low as $2,000. Another issue is mobility. Immersive systems tether you with cables to a computer. The effect isn't conducive to comfortable use over long periods. A third challenge is overall performance. To create smooth, slick virtual worlds, we must reduce system delays resulting from tracking systems and computation time. Delays affect sensory feedback, especially visual. It will be a long time before data goggle graphics offer the look and feel of Simulation VR and Projection VR.

We cannot overlook the significance of physiological issues in virtual reality. Researchers have yet to determine the impact of VR experiences on the human body and emotions. We do know that "simulator sickness" is not uncommon; headaches and nausea can accompany extended VR experiences.

Finally, people are just starting to focus on ethics issues. Who will be responsible for monitoring the accuracy of information presented in a public VR application?

Despite obstacles, VR's future looks exciting. We can expect the emergence of "networked virtual worlds" that will interconnect people in different locations with the help of phone lines. Multi-user capabilities promise great benefits; a world can be a lonely place if you're the only one in it.

Soon we'll be able to select from many low-cost virtual world tools as easily as we buy accessories for personal computers. We can expect to see better-looking virtual worlds, perhaps with photorealistic components. On the "high end," virtual worlds will incorporate another sense: touch. With special data gloves, we'll be able to feel virtual objects, sense their shape, texture and other physical attributes. And interaction with virtual worlds will one day include speech input and voice recognition.

If we compare the timeline of virtual reality's history to the timelines of aerospace and motion picture history, we see that we're only at the point in development equivalent to the Wright Brothers and Thomas Edison. As computer components get faster, less expensive, and more powerful, so too will virtual reality.

See you in cyberspace!

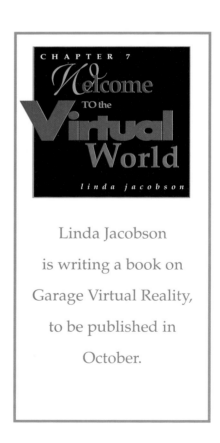

Linda Jacobson is writing a book on Garage Virtual Reality, to be published in October.

CHAPTER 8
Wonders of Water and Waves

peter sorensen

In the computer simulation of realistic images, it is relatively easy to synthesize man-made objects such as buildings, cars, and chrome logos, but the natural world is another matter. Here there is a level of complexity, generally without the straightforward underlying geometric structure found in manufactured objects. One of the most elusive natural phenomenon (for computer animation especially) is water—both for the way it moves (or in still pictures, the shape of the waves), and the way it optically affects light bouncing off it and passing through it.

Computer programmers love a challenge, and if they have a love of nature and an artistic bent, they are likely to be drawn to such elusive problems as animating water. They study the physics of fluids, optical distortion, the math, and so on. They even spend hours at the beach just meditating on the real thing.

As with all photorealistic computer graphics, there are two basic issues involved: modeling and rendering. Modeling in this case means creating algorithms that generate the shape of waves and how they change over time based on an approximation of fluid dynamics. Rendering tackles the issues of light transmission and thus how the final image looks.

> **modeling: the creation of a computer version of a real-world object.**

MODELING

It's not at all obvious how waves take shape and evolve the way they do. At first glance it would appear that waves are traveling masses of water, but this is essentially an illusion. All that's really moving is energy. If you observe a cork floating on the ocean some distance from shore you'll see that it moves up and down, and back and forth, but the cork basically just goes around in a circle with every passing wave and never gets anywhere as a result of the waves themselves (wind or current being separate considerations).

There are three kinds of waves. The biggest are the (inappropriately named) *tidal* waves, actually caused by geological forces such as earthquakes. The smallest are capillary waves, caused by wind, which can be just a fraction of an inch, and in between are those you usually see lapping at the seashore or lakeside. These generic waves get their start as capillary waves. When the friction of a good breeze ruffles a body of water (like a violin bow coaxing a note from a string), energy accumulates, forming ever larger wavelets.

> **rendering: using light and shadow to make a computer-generated model look realistic.**

These wavelets continue to grow, and soon they are big enough so they act like little sails, capturing even more of the energy from the restless air. Once started, the energy in the waves can travel great distances in the form of a "swell." By then the waves have formed into orderly "trains" that—oddly enough—progress at half the speed of the waves themselves. The lead waves gradually die out, each only getting half a wavelength ahead of its predecessor. But little of their power is lost. It only has been transferred down into the still depths, where it creates a reciprocating foundation for the following waves to ride upon.

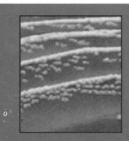

Beyond the sheer intellectual and artistic question, there are financial incentives that drive all computer graphics: the best software sells best. Realism in computer imagery is now expected by the public, being able to do realistic water is key (see Figure 8.1). The advertising industry has long been pushing for realism, and recently Hollywood has jumped on the bandwagon, so the competition among software providers is heating up in all areas. Additionally, in pure scientific research, there is a need to be able to visualize natural phenomena, and although water (and fluids in general) isn't the first thing on the list of models that scientists need, it is on the list.

Wonders of Water and Waves

Figure 8.1. *Surf at sunset with breaking waves. Created at Pixar in 1986 by Dr. Alain Fournier and William T. Reeves.* © 1986 Pixar.

When at last, after a voyage that perhaps has crossed the entire Pacific, the wondrous pulses of energy finally come to the shorebreak. Their final moments are the most interesting (and most complicated to animate). Why, at their end, do waves suddenly defy gravity and rear up like angry dragons attacking the land?

Remember two things: the water isn't really going anywhere (energy is) and a lot of that energy is hidden down in the depths of the wave train, as mentioned earlier. This invisible sub-surface energy is the key. Just as the cork was observed to be moving in circles on the surface, points below are also going around and around in circles of their own, called orbitals. The further down the orbitals are, the smaller they are. As long as the water is deep everything goes like clockwork and the energy is passed without fuss along the path of the wave train.

But when the waves reach the shallows, their orbitals become restricted, confined to ever tighter quarters, and the lower orbitals react by changing from circles into ellipses. This has the effect of slowing the forward progress of the wave, shortening its length, and making it taller. Finally, when the water becomes so shallow that its depth is little more than the distance between the waves, there is no longer enough water in front to satisfy the need of the ever-more compressed—and more energetic—

> orbitals: circular paths in which energy travels.

orbitals. The orbitals literally suck the water in from the front of the wave while pushing it from behind and squeezing the peak out into a lip. The leading edge of the lip pinches together, focusing its remaining energy so acutely that the water can reach escape velocity and shoot forth jets of foam and spray. After it plunges to its death, the resulting chaotic white water—only a ghost of the organized wave it once was—surges briefly up onto the beach before it returns to the sea like a soul to its reincarnation.

Along with the other exciting advances in computer animation in recent years, the art of simulating water has improved to the point where the line between actual water and computer-generated water can be very difficult to detect. A pair of researchers at Apple, Michael Kass and Gavin Miller, have devised a way of animating waves that is fast enough to be commercially viable. The method also models the previously unattained simulation of water "transport" (such as spilling over an obstacle and filling a depression on the other side).

At London's Digital Pictures, Mark Watt devised a technique for rendering "caustic" optical effects (those shimmering patterns on the bottom of a pool—see Figures 8.2 and 8.3) by tracing light rays in the same direction as nature (most "ray-tracing" software works backwards from nature) while combining other sophisticated lighting effects. And in Hiroshima, a team of computer scientists working with Dr. Eihachiro Nakamae developed software for automotive driving simulators with several significant advances, including the simulation of water on roads under various weather and lighting conditions. (See Figures 8.4 and 8.5.)

These people are following in the footsteps of such major CG water pioneers as Turner Whitted, Nelson Max, Darwyn Peachey, and Dr. Alain Fournier with William Reeves. It was Fournier & Reeves who, centered at Pixar in 1986, created a 10-second sequence of surf at sunset that had the first animated breaking waves (and with its rhythmically pulsating sea of liquid gold, it remains unsurpassed for breathtaking beauty—see Figures 8.1 and 8.6). Fornier (who concentrated on the problem of modeling, while Reeves was mainly concerned with rendering) recalls that, "We got started because I was looking at [physicist] Richard Feynman's notes on waves—he has a thing about waves—and his orbital model struck me as being very good for computer graphics."

ray tracing: using a computer to trace light rays in the same direction nature does in order to illuminate a computer-generated object.

Being faithful to the fluid mechanics of orbitals, this approach automatically relates the depth of the water to the wavelength, (which makes the waves diffract and follow the contour of the bottom properly) and, of course, causes them to break correctly when the water becomes shallow. Reeves recalls that they "split the water into three regions that were sort of glued together. There were the waves that were right up front, where we wanted as much detail as we could have, then there was a sort of middle ground where it wasn't as important to get so much detail, and then there was the stuff at the back where you almost couldn't see wave crests even—they were there in the model and in the data, but that was probably overkill."

Wonders of Water and Waves

Figures 8.2 and 8.3
Caustic light effects on the surface and bottom of a swimming pool. By Mark Watt at Digital Pictures using backward beam tracing. © 1990 Digital Pictures.

It was a simple matter to model the ocean at different resolutions, the trick was gluing the pieces together so the seams wouldn't show. Reeves says he took a chance and just "left a small gap between them with no water (as generated by the model) and fitted a flexible surface in between that stretched from one piece to the other. It didn't follow any oceanic principles at all but it was so small—just a couple of feet—that I don't think anybody noticed anything."

Fournier was concerned with the waves themselves, and not the transport of water from one place to another, but Kass has taken an entirely different tack with his modeling technique in order to economically deal with this problem. (Miller, his partner, focused on the rendering.) Ignoring orbitals altogether, Kass uses a "height field" to visually approximate what's going on. First he divides the volume of water up into a grid of vertical columns. Based on its height from the bottom, each column contains a known volume of water. It

On the Cutting Edge of Technology

Figure 8.4. *Wet road surface at night rendered using multiple layers of texture maps to create the puddles and their reflections. Directed by Eihachiro Nakamae at Hiroshima University. © 1990 Hiroshima University.*

Figure 8.5.
Wet road surface in the evening. Directed by Eihachiro Nakamae at Hiroshima University. © 1990 Hiroshima University.

Figure 8.6.
A beach seen from above with water being absorbed by the sand. Dr. Alain Fournier and William T. Reeves. © 1986 Pixar.

also has a known direction and velocity associated with it. From frame to frame, as the force of gravity tries to pull the higher columns down, the volume of water is conserved by being passed on in the proper direction and at the proper speed to the neighboring columns.

Compensating for the lack of orbitals with a fudge factor, Kass treats each column as if the whole thing were moving at the same speed (in nature the deeper water moves slower). This might horrify an oceanographer, but it looks fine on the screen. And it's so fast that it can model 1,032 columns of water in real time. "Our goal was to find the simplest approximation to the physics that would give realistic looking results so it would be possible to use it in a practical animation system," Kass says.

Recently, programmers Carl Ludwig and Dr. Eugene Troubetskoy (the father of ray tracing—who was computing rays of atomic particles before anyone dreamed of applying the principle to graphics) at Blue Sky Productions, devised an efficient algorithm for creating blobby water droplets, drooling honey and melted chocolate (see Figure 8.7). (For another Carl Ludwig rendering, see Figure 8.7A.) Their fluid simulation makes use of unique "primitive" ellipsoid shapes that have fields of influence around them that respond to other nearby primitives to simulate surface tension. When two such ellipsoids are pulled apart, they act as though they were both contained within a rubber membrane that sucks in as the gap widens, until it breaks. Drops colliding with a rigid surface shatter into smaller

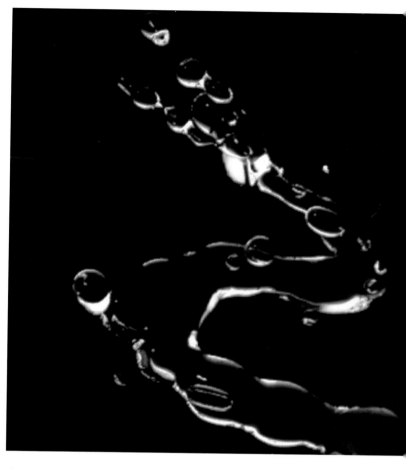

Figure 8.7.
Blobby water droplets modeled and raytraced by Carl Ludwig and Dr. Eugiene Troubetskoy (the father of ray tracing) at Blue Sky Productions. Their fluid simulation makes use of unique "primitive" ellipsoid shapes, which have fields of influence around them that respond to other nearby primitives to simulate surface tension. © 1980 Blue Sky Productions.

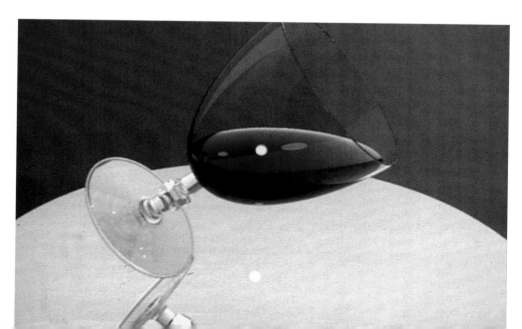

Figure 8.7A.
Ultra-realistic wineglass rendered with the world's best ray-tracing software, written by Carl Ludwig and his team at Blue Sky Productions. © 1980 Blue Sky Productions.

droplets. And when two different blobs touch, they adhere to each other and quickly pull themselves into a single blob. The major variable affecting the blob's behavior is its viscosity, which determines if it will react quickly like water, or slowly, like molasses in January.

Rendering

The way the sun shimmers off the dancing ocean waves and plays surrealistic games of light and dark down on the bottom of a pool is as beautiful and captivating as any natural phenomenon—right up there with the ethereal complexity of clouds and the flickering fury of fire. When it comes to rendering these patterns with computer graphics, water is every bit as difficult for us to portray as clouds and fire.

Once the water has been modeled using one of the approaches discussed earlier, a fair job of rendering its surface can be done with simple specular highlights, as if it were glass. We've seen "good-enough" water like that any number of times in TV commercials and print ads. But if you want people to really suspend disbelief and feel like they could jump in and swim, you have to render those elusive abstract light patterns that endlessly writhe on the bottom and/or reflect up onto the hull of a boat. Researchers call these magical effects *caustics*. (The term stems from the burning hot spot produced by a magnifying glass in sunlight.)

Logically, you would use ray tracing to render caustics, but no ordinary program is capable of doing the job. That's because standard ray tracing programs shoot rays backwards from the eye out into the scene, while the light focusing/scattering effect of caustics demands that the rays go in the proper direction if they are to be calculated with any practical speed whatsoever. Ray tracers usually trace backwards so they have to compute only those rays that actually enter the eye. But if you go from the eye to the bottom of a pool, you'd have to branch out from there in all directions to find out whether any wave anywhere was directing light to that spot. Multiply that task by all the rays you fire into the water and you realize the enormity of the problem. The trick is to figure out an efficient way to map the pool bottom with forward rays from the light source before firing your backwards rays into the scene. Within the past two years, researchers in Great Britain, Japan, and the United States have devised interesting ways of doing this.

Let's study the caustic refraction/reflection phenomena more closely for a moment. A wavy liquid surface functions essentially like a sloppy, irregular lens. When light passes through a convex (outwardly curving) lens, like an ordinary magnifying glass, the rays converge to a bright dot (called the focal point). If they go beyond that point, they start to diverge and the light energy diminishes. The less commonly seen concave (inwardly curved) lens immediately diverges the rays passing through it, so they never come to a focus.

Curved mirrors have a similar but opposite effect. An inwardly curved reflective surface will bring light to a focus, whereas an outwardly curved one will scatter rays away from each other. So this—with the added complexity that waves are irregularly shaped—is what causes both transmitted and reflected caustics. And it means that practically every ray striking a choppy surface bends or bounces off in a very different direction. Other than liquids, these things also happen with materials such as bathroom glass and crumpled aluminum foil.

Researchers who have recently been making images with caustic effects in them have all been employing variations of a two-step process involving ray tracing whole beams of light in reverse and then rendering the scene

> **caustics:** the light patterns created when light reflects and refracts through water.

> **specular:** a CG term referring to the directly affected hot spot on a shiny surface.

> **focal point:** the spot where the rays passing through a convex lens meet.

with conventional ray tracing in the usual direction. (Let me make it clear that I'll be using the term "backward" to mean the opposite direction from the way we usually trace rays in computer graphics. Even though backwards is actually forwards as far as Mother Nature is concerned!) Mark Watt of Digital Pictures in London calls this "backwards beam tracing" in honor of a 1986 paper by James Arvo on backwards ray tracing. But history buffs take note: the very first ray tracing paper in computer graphics, by IBM's Art Appel in 1963, discussed how he first tried tracing from the light source but then discovered the advantages of starting from the eye. Rendering guru James Kajiya recalls that, "It was a pioneering paper, way too far ahead of its time."

The problem, once again, is that caustic hot spots (and dark areas) are caused by varying densities of diffuse light that can be coming (or not coming) from anywhere on the water surface. ("The problem with diffuse light," says Kajiya, "is it comes from anywhere and goes to everywhere.") If you follow a ray out from the eye and strike an object affected by caustics, that spot could be lit by light directed from any part of the water surface, so there's no way that a single ray can tell how bright the spot is. You'd have to project a whole raster of rays out from the spot to do the job. And you'd have to do that for every spot affected by the caustics. A single frame could take as long to compute as an entire movie—forget it!

So, using a strategy somewhat like that used to render scenes with "radiosity" (very realistic diffuse lighting), some researchers first generate an "illumination map," like a texture map, for every surface affected by the caustic lighting. Generating the illumination map is the backwards part. You can imagine how you could fire out a grid of rays from the light source at the water, reflecting the rays up onto surfaces above, diffracting them down onto surfaces below, and thereby determining the patterns of light/dark for the maps.

That might be all there was to making the maps except that the dark areas caused by the caustics present another little problem: when the rays get sparse, they create spotty artifacts, and this spottiness can become nearly infinitely thin—just a dot all by itself without any neighbors within several feet. You'd really have to fire millions of rays from your light source and then gently filter the pattern for a nice result—and then your computation overhead would be going through the roof. Wouldn't it be nice if there were a way to not only avoid firing millions of rays, but to avoid those spotty rays altogether? This is where beam tracing comes in.

> **pencil: an optical term referring to a small bundle of rays.**

Mikio Shinya at Nippon Telegraph and Telephone Corporation has made some very realistic pictures using the illumination map process described earlier. He calls his process "grid-pencil-tracing." (Pencil is an optical term referring to a small bundle of rays.) He traces a grid of 128 X 128 pyramid shaped beams backwards from a point light source, through the surface of the water and on down to the bottom of a pool. When the beam encounters the water, the curvature and angle of the surface will diffract it accordingly. The beam spreads out if the surface is concave, concentrates it if it's convex, and redirects the whole beam this way or that depending on the overall angle. (Reflections do just the opposite, of course, but otherwise are treated identically.)

Once a beam strikes an object, the size and shape of the beam's footprint, (which Watt calls a "caustic polygon") paints the illuminance map with a shape having the appropriate brightness. If more than one beam strikes the same surface, the light accumulates to create realistic hot spots. Using a shading technique, the edges of the caustic polygons are easily smoothed out so they don't show.

One of Shinya's most effective pictures shows the caustic focusing of light after having passed through a glass of water and onto a table (see Figure 8.8). The softly feathered caustic is the result of light from a large rectangular source, as can be found on the ceiling of an office. Instead of projecting all the pyramid

shaped beams from a single point, the beams emanate from a grid of 7 X 7 locations spread across the rectangle, each of which emits many beams. Shinya has also modified his pencil beam process to economically simulate chromatic dispersion effects (the way a prism creates rainbows or a diamond flashes different colors) and cross-screen filters (as used by photographers to produce stars on highlights).

Figure 8.8.
Glass of water by Mikio Shinya at Nippon Telegraph & Telephone, using a modification of his "pencil tracing" technique to achieve a soft-penumbra effect from the light that has passed through the glass. © 1990 by NTT.

Mark Watt has come up with a clever variation of the backwards beam process that elegantly eliminates projecting any beams away from the water where they aren't needed. Instead of starting with a whole grid of beams that emanates from the light source in the approximate direction of the water, he first connects the vertices of the polygons forming the water surface to the light to make his beams—thus every poly counts. He has found that the water polygons need to be much smaller for this purpose than would normally be required for simple specular rendering.

Last, but perhaps most peculiar, a real smoke and mirrors way to render waves first devised in 1984, and recently revived in a more streamlined form: The smoke is simply "depth cuing" (*fog* to the uninitiated). But the mirrors—mirror singular, actually—is one vast, shiny polygon, flat as a pancake, representing the ocean (see Figure 8.9). Carl Ludwig, Ken Perlin, and Josh Pines at MAGI/Synthavision (the very first computer animation production company—now gone) used multiple cycloids to procedurally generate a reflection map with "perturbed normals" for the ocean. (A "normal" is what controls the angle that light bounces off of a simulated surface—"perturbing" the normal changes the angle.) So, instead of modeling their waves as actual 3-dimensional objects, the MAGI people translated the waves into regions of different reflective angles on the big, flat poly, and—as long as you didn't get too close— the result looked great. In fact, it was so realistic, Pines recalls, that at SIGGRAPH that year, "People actually voiced complaints that MAGI shouldn't have used stock water footage in their demo reel!"

Very sophisticated texture maps were among the tools developed in 1990 by a team of computer scientists working with Eihachiro Nakamae at Hiroshima University to render wet streets for an automobile simulator. They made extensive studies of road surfaces and puddles under the full range of moisture conditions and at all times of day and night (refer back to Figures 8.4 and 8.5). They have reproduced the look of asphalt under these conditions with many layers of texture, reflection and bump maps. The puddles, for instance, can have maps that simulate "mud particles that are considerably larger than the wavelength of light," that are perfectly diffuse and spherical (see SIGGRAPH 1990 proceedings for the math!). Their demonstration animation also shows soft shadows with penumbras, and diffraction effects that would be caused by scratches on the windshield and even by a driver's eyelashes.

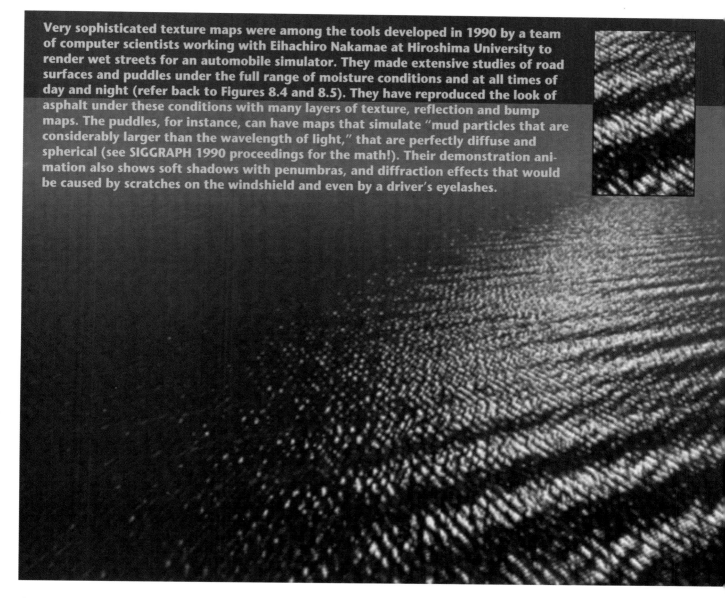

Figure 8.9.
Ocean surface created by Carl Ludwig, Ken Perlin, and Josh Pines at MAGI/Synthavision in 1984 still looks good today. It is actually a perfectly flat surface rendered with the normals perturbed (reflective angles altered) using an algorithm based on oceanic principles. © 1984 by MAGI.

Today Ludwig is at Blue Sky Productions where he recently resurrected his slight-of-hand method to animate water around dolphins leaping in a moonlit scene. But he wants it known that, "We certainly are capable of producing 3-D waves if we need to, but it's the old story of using whatever works. And from any significant angle above the water, perturbed normals are perfectly adequate."

As an old surf bum, I must say I'm very pleased with the advances that have been achieved in portraying water and waves with computer graphics. I think there needs to be work done on the subtleties, like spray, foam, and the uneven brown sandy patches stirred up by rip-tides. Then we can have a virtual reality experience where you stand on a hydraulically activated board, put on the stereoscopic headset, and surf the infamous "Banzai Pipeline" in perfect safety. Actually, writing the computer program would be more fun—who wants to surf the Pipeline "in perfect safety?!"

Chapter 8
Wonders of water and waves
peter sorensen

Peter Sorensen is an editor with *Computer Graphics World*.

CHAPTER 9
Artificial LIFE

john iovine

Creating artificial life-forms is one of the most interesting things you can do with a computer. Artificial-life programs also are called Cellular Automation (CA) programs and Genetic Algorithms (GA). Whatever they are called, this form of life is electronic and exists only within the confines of the computer system.

> **artificial life:** electronic life-forms created by and confined to computer systems.

Artificial-life programs are not merely toys for the technically inclined. They are used to successfully model biological organisms and ecosystems. Cellular automations can mimic evolution, co-evolution, the migration patterns of birds, colonies of ants, social order of bees, and growth of bacteria colonies. If that isn't enough, chaos algorithms can be incorporated into the programs to add a measure of randomness. The randomness factor allows accurate modeling of weather patterns, population growth, and the spread of infectious disease in populations. Artificial-life programs are being developed to help optimize neural networks, artificial intelligence, and parallel processors. Computer experiments are underway trying to structure CA programs to create and wire neural networks patterns. When these programs become successful, they essentially will learn by themselves.

LIFE

No discussion of artificial life would be complete without mentioning John von Neumann, a Hungarian mathematician, whose seminal work in the late 1940s on self-replicating programs originated cellular automata. Martin Gardner's "Mathematical Games" column in *Scientific American* brought cellular automation to the masses. The "Game of Life" originally was described in *Scientific American* in the early 1970s by Martin Gardner. John Horton Conway, a British mathematician then at the University of Cambridge in England, created the game. Professor Conway currently is teaching at Princeton University.

The Game of Life is played on a large two-dimensional grid of square cells, similar to a checkerboard. Each cell or block of the checkerboard can contain an organism, identified as a dot. Each cell on the checkerboard is surrounded by eight neighboring cells. Four cells are adjacent orthogonally and four cells adjacent diagonally (see Figures 9.1 and 9.2).

In the Game of Life, an organism dies or reproduces according to a few simple rules. In deriving rules, Conway sought to find the simplest rules that would create populations with the largest diversity and unpredictability. The rules to play one game of Life are as follows:

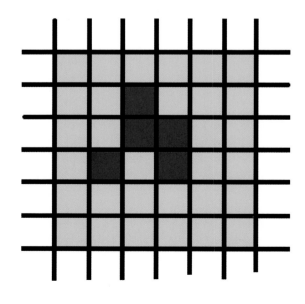 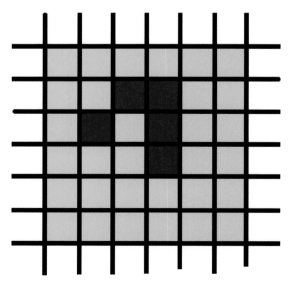

Figures 9.1 and 9.2. *Two consecutive generations from a Game of Life.*

1. **Survival.** Every organism with two or three neighboring cells that also contain organisms survives to the next generation.

2. **Death.** Every organism with four or more neighboring organisms dies of overpopulation. Every organism with just one neighbor or no neighbors dies of isolation.

3. **Reproduction.** Any empty cell next to exactly three neighbors is a birth cell. An organism is placed on it in the following generation.

To start the game, an initial pattern of live cells is placed in the grid. The colony of cells might grow into a large population, fall into a cyclic pattern, or die off. The complexity and diversity of the life generated in the game exceeded what anyone ever thought the underlying simplistic rules would create.

3-D Life

The Game of Life has advanced beyond a two-dimensional game. Carter Bays, a computer scientist at the University of South Carolina, has created three-dimensional versions of life. One version uses cubes held in three-dimensional space. A cube in 3-D Life has 26 neighbors instead of eight. The other 3-D version uses spheres. Unfortunately, I don't have the space to explore these programs in the chapter of this book. Don't worry; if you're interested in either version of 3-D Life, you can receive more information by contacting Carter Bays at the University of South Carolina:

> Carter Bays
> Computer Science Department
> The University of South Carolina
> Columbia, SC 29208

Boids is a computer-animated program of bird creatures created by Craig Reynolds of Symbolics Inc. The computer birds flock like real birds flock. Reynolds accomplished this with three simple rules. Maintain a minimum distance from the nearest object, match velocity with the nearby flock, and fly toward the greatest concentration of the flock.

Ants is a computer-animated program of ants. Created by David Jefferson and Robert Collins at UCLA, the computer-generated ants form colonies and have developed the ability to navigate electronic mazes in their search for symbols that represent food.

Belousov-Zhabotinskii

Self-organizing artificial-life programs exhibit evolving forms and structures. The programs also can mimic what is known as the Belousov-Zhabotinskii reaction. This is a simple chemical experiment undertaken approximately 30 years ago in Russia. It led to the discovery and exploration of self-organizing systems.

In the now-classic experiment, a solution of bromate ions in a highly acidic medium sits in a shallow dish and spontaneously forms centers of chemical activity. Each center creates expanding patterns of concentric, circular rings and spirals. As the patterns grow, they begin to crash into neighboring patterns. Soon the entire surface of the material is organized into a unique layout. The design then begins to decay and disappear as secondary reactions impede the primary one.

The experiment's visual impact 30 years ago was an unexpected but natural phenomenon. Today, the study of self-organizing reactions goes beyond its roots in chemistry into disciplines such as physics, cosmology, and biology. Many experts are convinced that self-organizing chemical reactions were a beginning stage in the development of life.

Genetic Algorithms

Genetic algorithms are a class of artificial-life programs. These programs evolve in a Darwinian (survival of the fittest) manner. I must specify Darwinian evolution because other researchers are busy using GA programs to test coevolution theories. Genetic algorithms are usually problem specific. They are designed to optimize their code for specific problem-solving tasks.

> genetic algorithms: a class of artificial-life programs that optimize their code for specific problem-solving tasks.

Genetic programs are unique in the fact that they can mate with other program code to produce "Combined Code" offspring. They also can mutate to generate new program code.

You might be curious as to how a program can mate with another program or mutate its own code. Actually, it's quite simple. Start with a population of individual program codes. At its most elementary level, all programs are strings of binary numbers. Let's assume this 8-bit binary number 00101001 represents a possible solution value in a genetic algorithm. The computer could mutate the value by switching the value of any bit in the number, for example 00101001 → 10101001. This new value would then run through the genetic algorithm. If it achieves better results than the original value, it would be saved; if not, it would be eliminated. Mating programs are just as easy. Two binary numbers crossover specific bit values and reproduce.

```
Bit Position    8 7 6 5 4 3 2 1

Initial A       1 0 1 1 1 1 0 1
Initial B       1 1 0 1 0 1 1 1

Final A         1 0 1 1 0 1 1 1
Final B         1 1 0 1 1 1 0 1
```

In this mating example, the binary numbers exchange binary values 1–4. So the final value of A has the same bit values for bit positions 8–5 as the Initial A and has the bit values of the Initial B in bit positions 1–4.

The new values are tested against the problem at hand, and measurements of their fitness or level of performance determine their fate. As before, if there is an increase in performance the new values are saved.

The language of genetic algorithms borrows many biological terms such as *mating*, *genes*, *reproduction*, *fitness*, and *survival*. Programmers consider the binary values of the program code electronic DNA.

Genetic algorithms are inherently parallel processes. Each processor operates or performs on a self-contained program code. There might be hundreds or thousands of program codes mutating, mating, and evolving toward the most efficient code possible.

Unlike nature's evolution, which is relatively slow, genetic algorithms mimic evolution at millions of instructions per second (MIPS).

Danny Hillis, the founder of Thinking Machines Corporation in Cambridge, Massachusetts, studies genetic algorithms. In one of Hillis' genetic-algorithm experiments, he required the use of one of his company's Connection Machine computers. This is a super computer that contains 65,536 parallel processors, allowing it to simulate 65,536 independent organisms quickly. The artificial organisms used in the experiment were numerical sorting programs. To mimic evolution, the programs could randomly change (mutate) their code. In addition, they also could combine their program codes with other program codes (organisms) in a process similar to artificial sex. All new programming codes created were tested for efficiency in sorting numbers. Only the fittest or most efficient programs survived in a manner similar to natural selection. This electronic evolution developed improvements in program code quickly.

Genetic algorithms will become more popular as parallel-processor computers become more accessible. Looking back at Hillis' experiment, if his programs were run on a standard serial computer, they would have taken at least 65,536 times longer to run.

Writing Music

Artificial-life programs can write music. Eric Iverson of New Mexico State University created an artificial-life program that creates new music from old music.

Most computerized music generation programs rely on generating a large number of random notes. The random notes then are filtered through an artificially intelligent (AI) subroutine. The subroutine section contains "rules" that constitute the programmer's best version of what sequence of notes can make music.

The artificial-life music generation program operates differently. Iverson wrote a music-metabolizing program called Metamuse. Essentially, the artificial-life program eats the

You might think genetic algorithms are somewhat esoteric. This is not the case. Genetic algorithms are being used at the General Electric Corporate Research and Development Center in Schenectady, New York. GE engineers were given the task of improving the fuel efficiency of a jet engine. Improvements by one or two percentage points in fuel efficiency will give their engines a more competitive edge.

David Powell, a GE computer analyst, and his colleagues were able to program a genetic algorithm to help in their engine design for a new Boeing 777. The genetic algorithm explored thousands of design options and combinations engineers normally don't have time to check.

The creation of the program called Engineous was a combined effort of the GE team and the science department at Rensselaer Polytechnic Institute in Troy, N.Y. The Engineous program is a hybrid containing both genetic algorithms and expert systems.

music fed to it. Metamuse first randomly extracts a string of four notes from the composition. That string serves as a digestive enzyme. The program searches for the same sequence of notes in the entire composition. When it finds the sequence, it cuts the music in two in the middle of that sequence. The enzyme is rewarded for finding a similar sequence by being allowed to reproduce itself. The two copies of the enzyme then continue the search for more matches. At the point the enzyme reproduces itself, there is a small chance it will mutate. The sequence mutates by changing one of its four notes. The digestive process ends when the composition has been reduced to a number of fragments approximately the size of the enzymes themselves.

Metamuse now operates in reverse. It begins to glue the fragments together. The program looks for two fragments that have the same note in the starting or ending position and glues them together. A music fragment DFAG might be linked with fragment AADD on one end and GFGF on the other, making the longer sequence AADDFAGFGF. Sequences are allowed to copy themselves with a small chance of mutation.

Metamuse tries to preserve patterns in the music while at the same time adding switches and changes. Because the program doesn't contain any rules on music composition, it is suitable for all forms of music. You could feed the program anything from Madonna to Mozart.

Iverson states that one-third of the time the program achieves something he really likes. The other two-thirds of the time, it's off in the ozone.

COOPERATIVE CO-EVOLUTION

Darwin's theory of evolution specifies survival of the fittest, not the nicest. Two Austrian mathematicians have discovered, through the use of cellular-automation programs, that cooperative and unselfish behavior might play a role in the story of evolution. Karl Sigmund is a 47-year-old mathematician at the University of Vienna. Martin Nowak is a former graduate student of Sigmund. Together, they discovered this revolutionary theory. In fact, their new theory is supported in nature. For instance, the vampire bat will share their blood meals with unrelated, less fortunate, neighbors. Vampire bats must consume 50 to 100 percent of their body weight in blood every night. If a bat fails to feed for two nights in a row, it will die. However, a bat can gain another 12 hours of life and another chance to feed if it is given a regurgitated blood meal by a roost mate. If the bats didn't practice food sharing, their annual mortality would be 82 percent. With food sharing, the number drops down to 24 percent. Obviously from an evolutionary standpoint, this unselfish behavior benefits the propagation of the species. Bats are not the only species to practice unselfish behavior.

Cellular-automation programs vary in levels of sophistication. Obviously, Hillis' CA program is far too complex to program onto simple serial computers. However, cellular-automation programs generally consist of a few simple rules. When the computer rapidly and consistently repeats the rules through generations, complex patterns can evolve that are not written into the program; they develop by themselves. The program's striking behavioral patterns provide windows of opportunity to study and to model living systems and artificial intelligence.

FREE ARTIFICIAL-LIFE PROGRAMS

After reading this chapter on artificial life, you might want to examine a few cellular-automation programs for yourself. If you belong to a large computer Bulletin Board System (BBS) or network such as CompuServe, there are artificial-life programs in various Special Interest Groups (SIGs) that you can download for free. I checked CompuServe's IBM File Finder, Go IBMFF, and found about 30 artificial-life programs I could download (see Figures 9.3, 9.4, and 9.5).

The most common artificial-life program is the classic version of "Life" by Conway. The game of life exists under multiple listings written with a variety of computer languages.

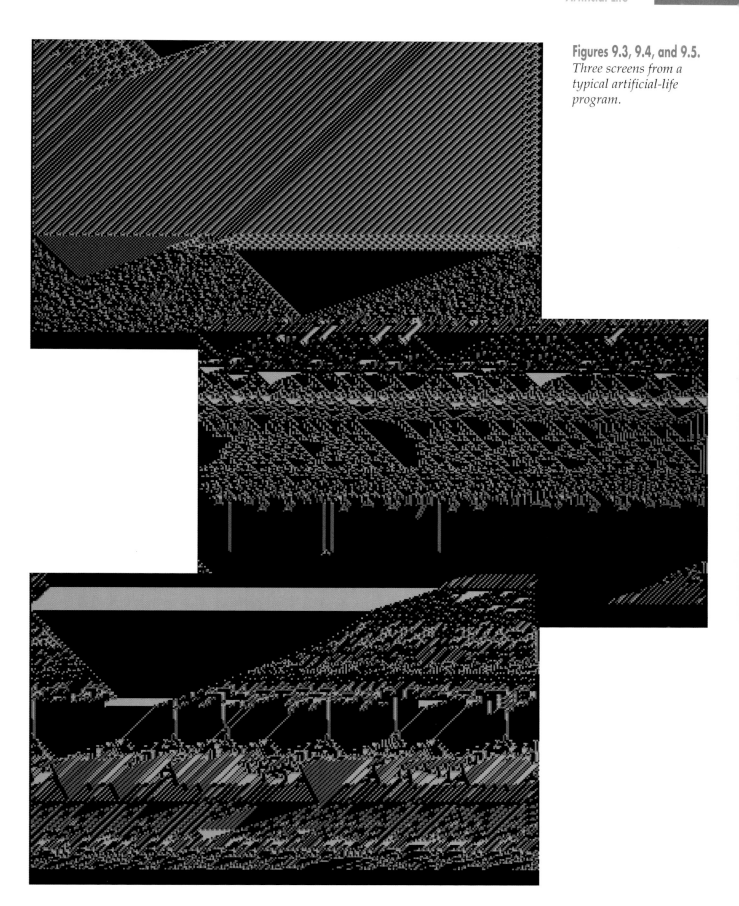

Figures 9.3, 9.4, and 9.5. *Three screens from a typical artificial-life program.*

On the Cutting Edge of Technology

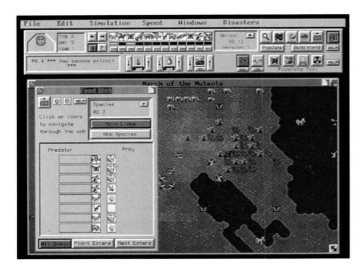

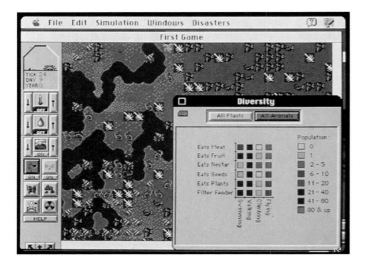

COMMERCIAL ARTIFICIAL-LIFE PROGRAMS

More sophisticated artificial-life programs are available commercially. These automation programs are available through standard distribution (stores) or directly from the manufacturer.

In SimLife, a biological simulation, you custom design the environment and life-forms (see Figures 9.6, 9.7, and 9.8). The best adapted species survives. Cellpro is a standard cellular-automation program that allows the user to adjust a large variety of parameters.

For the Amiga	For the IBM & Amiga
CellPro	Simlife
MegaGem	Maxis
1903 Adria	2 Theater Sq.
Santa Maria, CA 93454	Suite 230
(805) 349-1104	Orinda, CA 94563
	800-336-2947
	510-254-9700

Maxis also has released El-Fish, an aquarium-generating program for your computer (see Figures 9.9, 9.10, and 9.11). You can use the software to create new species of fish by mating existing species (see Figure 9.12).

Figures 9.6, 9.7, and 9.8.
Three scenes from the Maxis game SimLife. (Images courtesy of Maxis.)

Artificial Life

THE DARK SIDE OF CELLULAR AUTOMATION

The dark side of cellular automation consists of computer worms and viruses. Computer worms and viruses are self-replicating programs that instigate themselves into computer operation systems. Viruses, like their biological counterpart, infect and replicate inside hosts. For computer viruses the hosts are program files and diskettes.

Figures 9.9, 9.10, and 9.11.
Three aquarium scenes from EL-Fish. (Images courtesy of Maxis.)

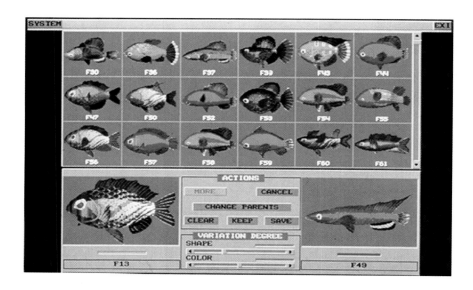

Figure 9.12.
With El-Fish, you can create your own species of fish. (Image courtesy of Maxis.)

computer worms and viruses: self-replicating programs that instigate themselves into computer operation systems.

Viruses are spread from computer to computer through "infected" diskettes or from computer networks. Viruses that hide in the boot section of the computer's hard disk or floppy diskette are called boot-sector viruses. From this vantage point the virus attempts to spread itself by infecting any disks it has the occasion to corrupt.

The "Stoned" virus is an example of a boot-sector virus. This virus destroys the directories and file allocation tables (FAT) of the hard drive while displaying the message "Your PC is stoned—LEGALIZE MARIJUANA." The file allocation table is a road map the computer writes on the disk telling it where it has stored all the files.

A file virus is another type of virus. These viruses can either infect or attach themselves to program files. If the virus infects a program, it writes viral code into the program itself, and that generally renders the program inoperative.

A virus that attaches itself to a program waits until that program is executed and loads itself into memory along with the program. For instance, if you ran a word-processing program with an attached virus, the virus would load itself into the computer memory with the word processor. Once it's in memory, it stays there even when you're finished using the word processor. The virus stays in memory waiting for another program to run. If you should run another program (such as a drawing program), the virus attaches itself to that program. This process continues until the virus has infected every executable file on your hard drive.

Worms are a class of viruses. These are usually found on mainframe computers. The most common attack of a worm is to replicate itself *ad infinitum* until the entire storage system of the computer is full of worm copies. As the worm eats up RAM and disk space, the computer slows down drastically.

Bombs are viruses that are triggered by a date, time, or event. The Jerusalem-B virus goes off every Friday the 13th. It's main attack is erasing any program run on that day.

Stealth viruses can be either boot or file viruses. They are called stealth viruses because they try to hide from antiviral programs. The manner by which the virus hides is ingenious: it changes the disk directory information to conceal its size and location.

Some computer scientists speculated that viruses are the first programs capable of existing without willful cooperation of humans. While this is true, it must be remembered that without humans to write the program code originally, viruses wouldn't exist at all.

Because of the potential damage that computer viruses can create, there's a market for antiviral programs. Microsoft has included antiviral software with DOS 6.0. Microsoft claims its software can detect and protect the user from more than 800 different types of viruses. There are many commercial antiviral programs on the market that prevent viruses from attacking your computer system.

The National Computer Security Association has an eight-page document, "Corporate Virus Prevention Policy," for sale. The document is targeted toward individuals who are responsible for their companies' antivirus policy. The document costs $9.95 for a printed copy, $12.95 for a disk with ASCII text file, or $19.95 in WordPerfect format. Call 717-258-1816 for more information.

CORE WARS

Core Wars are a recreational form of computer viruses. The game was created by A. K. Dewdney. The game is simple. Two programs are placed into a computer's memory. Each program's job is to battle and annihilate the other program. The winning program takes over the allocated space in the computer's memory.

The initial publication of the game was in the May 1984 issue of *Scientific American*'s "Computer Recreation" column by Dewdney.

Since the initial description, the game gained so much popularity that tournaments are held for the best Core War program.

Further information on Core Wars can be received by writing:

The Core War Newsletter
William R. Buckley, Editor
5712 Kern Drive
Huntington Beach, CA 92649

International Core War Society
Mark Clarkson
8619 Wassall Street
Wichita, KS 67210

Software available from:

Amiga PCs
Mark A. Durham
5712 Kern Drive
Houston, TX 77054

IBM PCs
William R. Buckley
8282 Cambridge Street #507
Huntington Beach, CA 92649

CHAPTER 9
Artificial Life
john iovine

John Iovine is currently writing a book of neural networks to be published by Sams in the fall

CHAPTER 10

Fractals

peter sorensen

Fractal research is arguably the most exciting cutting edge of mathematics today. Many people have heard about fractals, but most are at a loss to explain what these mysterious things are—beyond the fact that they have something to do with chaos theory and psychedelic graphics. Yet the concepts behind them are simple (although, admittedly "simple" and "new" together in one sentence can be a problem).

What we're dealing with is a new kind of mathematics that probes the heart of natural phenomena—where no one has gone before. The uses for fractals range widely, through math and physics, biology and sociology, to art and even motion picture scene simulation.

There are two keys to understanding fractals: "self-similarity" and "fractal dimension." The concept of self-similarity is easy: a fractal looks very much the same whether you are looking at it from far away or if you are looking at a small detail close up. A lightning bolt, like much of nature, is a fractal, and is an excellent example of self-similarity. The whole bolt is a jagged line. But look at a photo of one with a magnifying glass, and you'll see little jagged sparks shooting off of it. If you could zoom in closer and closer, you would find ever smaller sparks branching off the parent sparks, and so on, almost ad infinitum. Clouds, with their big puffs made of smaller and smaller puffs, are another example of fractals in nature.

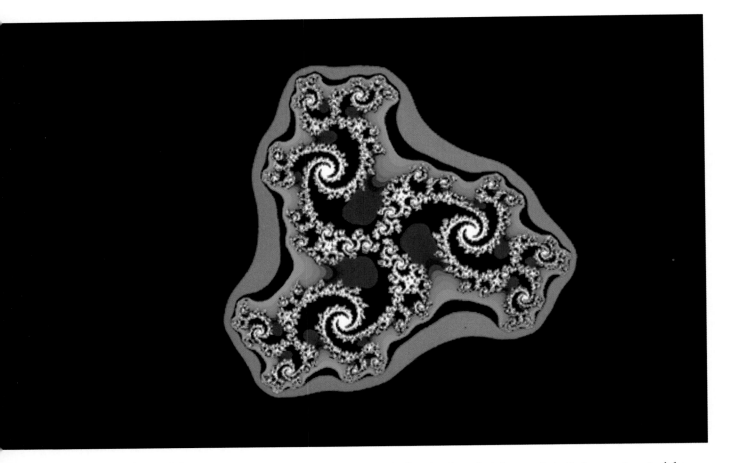

Figure 10.1. *Fractal spirals with domains of color indicating how fast those points that are not part of the fractal itself are being pushed away from their point of origin by the generating function. The red is the fastest and the blue is the slowest. The fractal was created by Greg Turk using a color program written by David Coons. (c) 1983 Greg Turk.*

Mathematically, fractals are infinitely complex, with no limit to their self-similar details, but in the real world they do have their limits. A tree is a good example. The branches have branchlets, which have twigs—but that only goes so far before they terminate in buds. A computer-generated fractal is limited only by the computer's memory capacity.

The second key ingredient, the fractal dimension, is the most important. It's even the derivation of the word fractal. Here's an example that demonstrates the concept: You know that a sheet of paper is two-dimensional (if you ignore the thickness). If you fold part of it, it enters 3-D space, but the surface itself is theoretically still just 2-D. That's the old way of thinking about it. In fractal math, we still don't count the 3-D space it folds into, but the crease itself is worth something—some tiny fraction of a dimension. Consider a piece of paper that's all crumpled. The more complex the wrinkles, the bigger the fractal dimension between 2 and 3. It would have to become infinitely wrinkled to actually reach 3. The fractal dimension is expressed as a decimal, such as 2.342.

The history of fractals began in the later part of the last century when a few maverick mathematicians began to rebel against the idealistic concepts that had persisted since the days of Euclid. The prevailing thought in those days was that the universe could be described in terms of the perfect forms of standard geometry and the dynamics of Newton. But nature isn't neat and tidy like that. "Clouds are not spheres, mountains are not cones, coastlines are not circles, and bark is not smooth, nor does lightning travel in a straight line," says Dr. Benoit B. Mandelbrot, in his seminal book, *The Fractal Geometry of Nature* published by W. H. Freeman and Company. The sixty-eight year old French genius who is today's foremost fractal pioneer—and who coined the term fractal—was inspired by those early mavericks who dared to go against the establishment (who considered the fractal theories "pathological," "psychotic," and even "terrifying"). Mandelbrot spent many thankless years on the problem before the establishment could bring itself to recognize the power of the concept.

Some of the more common fractals are Julia sets (see Figures 10.2 and 10.3), convoluted forms whose twists and turns remind one of the scales on a sinuous dragon. The most commonly seen fractal is the Mandelbrot set, named after (but not by) "the good doctor," as Arthur C. Clarke calls him. The M-set is actually the sum of all the Julia sets (of which there are an infinite number—each of which is infinitely complex!). Other fractals can be made to look exactly like mountain ranges (see Figure 10.4), or plants (see Figure 10.5), or even to control the motion of complex animated scenes. Perhaps the most fascinating fractals are those in the fourth dimension, with bizarre and beautiful organic contours of which we can only see 3-D "shadows."

Whatever form they take, fractals would be impossible for us to create, for all practical purposes, without the aid of computers (see Figure 10.6). Although the formulas that generate fractals are simple enough, they must be repeated over and over—each time using the result of the previous calculation as the start of the next—many millions of times to fill a computer screen with color.

> self-similarity: when an object looks the same whether it is examined close-up or from afar.

> fractal dimension: The extra dimension an object takes on when it isn't completely in solid dimensions.

Figures 10.2 and 10.3. *Colorful spiral detail of a Julia set, and a spark-like detail of the Mandelbrot set, by fractal artist Charles Fitch. © 1988 by Charles Fitch.*

Fractals 109

Figure 10.4.
Classic fractal mountains created by pioneer Loren Carpenter in 1979. Carpenter's work proved that one day computer graphics would be an integral part of the entertainment business. © 1979 Loren Carpenter.

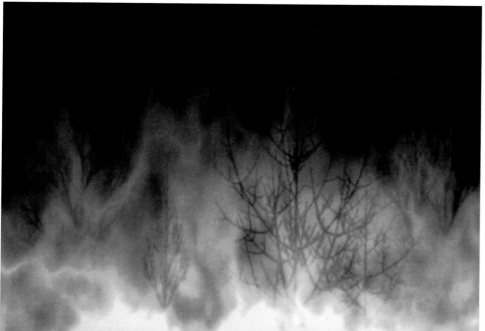

Figure 10.5.
Fractal forest fire simulated by Ken Musgrave, C.E. Kolb, L. Emme, and Dr. Benoit Mandelbrot after "studying lots of lighted matches in the office," Musgrave says. The trees were grown using L-systems, a technique similar to fractals, in a nature-mimicking program by Dr. Przemyslaw Prusinkiewicz, a computer-science professor at the University of Regina (saskatchewan, Canada). (c) 1990 Musgrave, Mandelbrot & Prusinkiewicz.

Figure 10.6. *Ray-traced fractal tetrahedron, hints at the infinite fine detail of its Swiss-cheese structure, in an image made by John Hart, then at the University of Illinois.* © 1990 John Hart.

Figure 10.7. *"Stalk" structures discovered by Dr. Clifford Pickover inside the Mandelbrot set, and not usually seen. Pickover is an eclectic computer scientist whose work on biomorphs, speech patterns, the shroud of Turin, seashells, and so on, always result in great graphics.* © 1989 Cliff Pickover, IBM.

In computer graphics there are two different categories of fractals: perfectly pure mathematical entities, such as the Mandelbrot set (see Figure 10.7), which are called deterministic, and the non-deterministic kind that have a random element thrown into the mix. This randomness is necessary for simulating natural forms, such as the branching of trees, the meandering of streams, and the awesome expanse of stars and galaxies throughout the heavens.

> **deterministic:** a purely mathematical fractal.
>
> **non-deterministic:** a fractal with some randomness to it.

Creating computer synthesized mountains (see Figure 10.4) involves generating a lot of random numbers to get the altitudes of the myriad peaks and valleys. If you wrote a program that just generated random numbers willy-nilly, you could wind up with your tallest peak right next to your deepest valley. Nature doesn't work that way—the peak of Everest isn't right next to Death Valley. Although you could tinker with your program so it took neighboring altitudes into account, you wouldn't be imitating nature properly unless you reinvented fractals to control the chaos.

If mountains are like a 2-D surface that's approaching 3-D, what about a 3-D surface that starts leaking into the 4th dimension? Einstein defines the fourth dimension as time, and as far as our existence in the physical world is concerned, that's true. But there are other ways of defining the fourth dimension, and the one we're going to explore is imaginary: A 90 degree angle "perpendicular: to our physical plane. That sounds crazy, but as far as mathematicians are concerned, the numbers can be made to work just fine and produce useful results, so let's take them at their word.

> A frequently used illustration of fractals in nature is the measurement of continental coastlines. Mandelbrot uses Britain as an example, observing that the ragged outline on a map has a fractal dimension somewhat greater than 1. If you look at maps with ever increasing detail the coastline's length seems to increase. "The result is most peculiar: coastline length turns out to be an elusive notion that slips between the fingers of one who wants to grasp it. All measurement methods ultimately lead to the conclusion that the coastline's length is very large and so ill-determined that it is best considered infinite." That's a fractal for you!

Graphically the technique of 4-D algebraic operations (called "Quaternions") produces fractals that are a sight to behold! (See Figure 10.8.) They have a peculiar déjà-vu quality, like you've seen them before—but not on Earth—which is a result of the fact that they are shaped by the fractal laws that have molded the environment in which we live. But they are alien as well, because their spawning ground is higher up. (Bear in mind that, since these strange things lurk in the fourth dimension, we can't look at them directly with our 3-space eyes, but we can use the computer to generate 3-D forms that are "shadows" of the beasts. For, just as your 3-D body casts a 2-D shadow on the ground, so a 4-D object can be made to cast a 3-D silhouette that can be rendered as a solid shape.)

Figure 10.8.
Mysterious quaternion, a four-dimensional generalization of a self-squared dragon by IBM's Alan Norton. This bone-like domain of attraction is modeled from 5.6 million points, and is only one of seven intertwining components that would fit together to comprise the entire 4-D dragon—a far more complicated structure. The formula for this quaternion is (1.475+0.906) X(1-X). (c) 1981 Alan Norton.

Meet Alan Norton, a computer scientist at IBM's Thomas J. Watson research center working on development of parallel processing supercomputers, who pioneered quaternion fractals using the research center's mighty array processors (see Figures 10.9 and 10.10). Norton says that the idea of dimension is, "sort of a continuum...and anything for which the dimension turns out to be somewhere on the continuum between whole integers would necessarily be a fractal. Not only do fractals have a funny dimension, but magnifying them will give you something that looks statistically the same—self similarity that is unvarying under the change of scale. That gives you something you can work with—and it turns out to be surprisingly effective."

Figures 10.9 and 10.10.
Two more quaternions by Alan Norton show the range of 3-D shapes that might be thought of as being slices cut at different angles through the four-dimensional fractal.
© *1989 Alan NOrton*

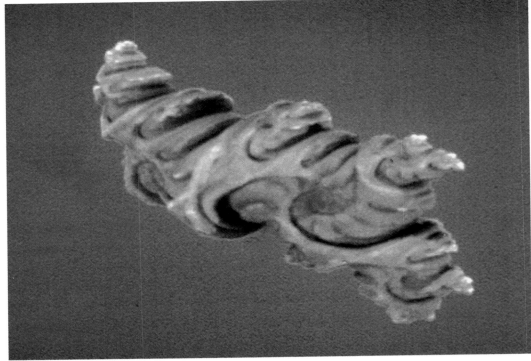

Why is it that nature is self-similar is the mystery. That nature is fractal is no mystery. Nature is fractal because nothing is perfectly smooth—a perfectly smooth surface is a mental abstraction. On the other hand not everything is self-similar, but the fact that it turns out to be so useful is the surprising thing that comes out of what Mandelbrot has been doing.

Norton started exploring quaternion fractals on a whim one day, "I was looking at some 2-D fractal dragon shapes and thought, 'I wonder what extension would they have to a higher dimension?' When I had the computer calculate different cross-sections of these things I found out that there was something interesting going on there."

Without being overly technical, he explains the process he uses to look for the quaternion shapes, which he calls "Domains of Attraction." "I apply the simplest sort of dynamical rules for four dimensions, and these rules cause points to be attracted to stable orbits or cycles, sort of like the way matter in the solar system was attracted to the planetary orbits. If you were in this universe where things don't behave according to the laws of gravity, but according to the simplest mathematics possible, then these are the sort of shapes that emerge in this world. They have different fractal dimensions, so one can make them look much more rugged or be relatively smooth. The simplest example would be the formula X squared, that would give me a sphere. Other parameters can be varied to provide recurring patterns in these shapes, for example repeating, looping strands and lots of interconnection relationships.

His method of actually making the picture once the quaternion shape has been calculated is different from most computer graphic techniques. Instead of using polygons he builds the shapes from tiny points, "like grains of sand," in space, storing their locations in memory. First he takes an educated guess as to where the dragon is in the picture and "fires a single ray" at it. The place where the ray strikes the beast becomes the starting point for the rest of the process. Instead of computing the remaining points with individual rays, which would be prohibitively time consuming, he has the program sort of crawl around on the surface of the object "like thousands of ants," feeling their way from point to point, checking to see if each point is inside or outside of the fractal.

Inspired by Norton, John Hart, then a student at the University of Illinois, started exploring the quaternions, and pushed the art to new heights. Hart's method of raytracing the fractals and rendering them is faster, and the two techniques result in subtly different aesthetics.

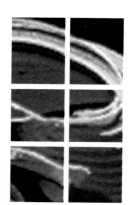

The fascinating forms of the quaternion fractals tap into the subconscious form-world of the mind, striking resonant chords—like seeing animals in clouds. This is what causes them to attract so much attention at computer graphics conferences and art shows. They look like plants and animals that we easily could imagine flourishing in the fertile ocean of some distant world. Musing on why this might be, Norton says, "The mysterious quality of these pictures comes from the fact that in some sense they exist. These objects have been lurking in abstract geometry and nobody's ever seen 'em before, but mathematically they exist just as much as a square exists and has an identity of it's own. You know, the ancient Greeks made a big deal about it when they discovered that the diagonal of a square is an irrational number—they thought that was such a remarkable thing that they sacrificed hundreds of cattle to the Gods and had a big party! Well, here is something that similarly is a true fact, but thanks to computer graphics it's something you can see with your eyes. I think there is something basic and essential about these shapes."

Although the quaternions are complicated, exploring fractals like the Mandelbrot set is something that anybody with even modest computer graphics capability can become involved with, as there are several such programs available now. The computer, like the microscope or the telescope, can make the invisible visible. It's ability to materialize mathematical abstractions—and far more than just fractals—will undoubtedly inspire a lot of people who would have otherwise been unable to appreciate such abstractions as raw textbook formulas. As Norton points out, "When you talk to mathematicians, and you ask them what it is about mathematics that interests them they say 'Well, mathematics is beautiful.' They perceive a beauty in these abstract structures which we usually can't see. And the thing about computer graphics is that you can actually take these structures and you can turn them around and see that they really are beautiful!"

Not only are they beautiful, but fractals rank among the most significant discoveries of our time. Fractals along with chaos theory, are revolutionizing our understanding of the natural world; helping us design quieter and cleaner engines; and even to get a better grasp on the functions of our own minds.

In the realm of the mind, clinical psychologist Dr. Terry Marks believes that fractals might be used as a metaphor to understand the psyche (the mental self). Of the fractal attribute of self-similarity, she says, "We each do the same things over and over, repeating the same patterns, repeating the same mistakes. We resemble ourselves over time in different situations with different actors. That's like the self-similarity of the buds on the Mandelbrot set. We can take any aspect of ourselves and look infinitely deep and find 'shapes' that look like the whole part."

When I called him to find out what was new to include in this book, Dr. Mandelbrot said he had just won the prestigious Wolf prize for physics, particularly for his application of fractals to physical phenomena such as DLAs.

Mandelbrot is into graphics as well, and often works with F. Kenton Musgrave on projects that produce very interesting computer images. Musgrave, who is both a programmer and an artist, has created exceptional pictures of fractal terrain, mountains, and lakes. The terrain work is of particular interest because of the accurate way it simulates natural erosion over time. At first the jagged fractal terrain looks like a geologically "young" surface, but then, as Mandelbrot put it, they "pour water on it and see what happens."

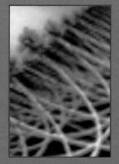

One rather unexpected use for fractals has been in ultra-high-ratio image compression for storing pictures on computer disks and transmitting video over telephone wires. Professor Michael Barnsley discovered that an image can be deconstructed in terms of its fractal content, and reconstructed just from the formulas of the fractals therein. His video modem can receive compressed frames and reconstruct them in real time, although the original compression takes some ten minutes per frame with current technology. Barnsley's fractal transform does impart a noticeable kind of unique digital graininess to high resolution images, but a small loss of quality would be acceptable for many applications in the light of the 500-to-1 compression ratio that can be achieved.

Dr. Mandelbrot's own recent work has been to intuitively apply fractal concepts to physics. He has been especially interested in the way cinders stick to one another, accumulating into little globs in a fractal manner, called Diffusion-Limited Aggregations (DLAs). "It's not too appetizing perhaps," Mandelbrot explains, "it's the way in which dirt forms in car engines and ashes form in chimneys. They swirl around and end up creating particles that are very complicated. Shapes of very great variety are formed which are a great challenge in mathematics and physics. They link many old and new questions of mathematics and physics together. They are quite amazing." And then he points out that, "most of the work in fractals is in physics, not in graphics. A big part of disorder in physics is fractal, so fractals are a very basic tool in physics now."

One of the more interesting discussions about fractals and nature centers on the question of whether the physical universe is one giant fractal or not. From the distribution of galaxies down to the sub-atomic scale we see fertile realms of self-similarity that have measurable fractal dimension. Mandelbrot, who says the fractal dimension of the largest cosmic features are between 1.23 and 1.3, declares that, "The big dispute about whether the large scale Universe is fractal or not is a matter of interpretation of data. Most people now believe that it is fractal over very large distances, but only down to the distribution of galaxies."

As with all natural fractals, the Universe breaks down at some point—like the branching of a tree stopping at the buds. As best we can tell, the largest-scale feature of the Universe resembles a sponge, with huge holes pushing the galaxies into the corners. It's a fair bet that the sponge is a fractal, as are the galactic clusters and the dusty distribution of stars within every galaxy. Shift your scale of view, and down on the surface of the Earth the shapes of rivers and trees, ferns and micro-organisms are classical fractals. And so on, down into the micro-structure of matter. It's as if nature makes gear-shifts now and then. Possibly these apparent discontinuities are due to our limited perception of some magnificent fractal of a very high dimension, the shadow of which is our four-dimensional space-time!

As Alan Norton discussed, when the ancients discovered geometry they literally worshiped the proportions of circles, triangles, polygons, and polyhedra. The structure of creation can be understood in terms of these things and their harmonic relationships. Now, after thousands of years we have discovered a totally new tool for understanding the nature of the Universe: fractal geometry. A few brave mathematicians secretly admit they feel there's something profound about the elegance of the Mandelbrot set—just like the circle and triangle. Maybe we should throw a big party!

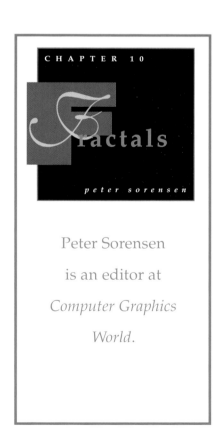

CHAPTER 10

Fractals

peter sorensen

Peter Sorensen

is an editor at

Computer Graphics World.

CHAPTER 11
Fuzzy Logic

dr. jack aldridge

When was the last time you used a precise value to make a decision? Would you really have decided not to buy that new car if its price had been $20,001 instead of $19,995? How about $20,100? Mostly, we use more qualitative ranges, such as "the price is high" and "that car looks very good" for decision making near the boundary between one decision and another because there is always some room for favoring multiple options there. Often, factors outside of direct sensor measurements enter the decision process. For example, if a room feels hot or muggy, you might decide to run the air conditioner, unless you were just leaving for the rest of the day. You might equally well decide to ignore the latter fact in making the decision.

Current control systems operate in the opposite mode. They measure a parameter such as temperature and slavishly turn the heater or air conditioner on or off when the temperature falls above or below the setting of the thermostat.

Such considerations of complex systems prompted Lotfi Zadeh to introduce the concept of fuzzy sets in 1965 in an article titled "Fuzzy Sets." This paper started the field of fuzzy logic. Fuzzy logic is a mathematically based scheme for modeling and controlling complex systems by recognizing that complexity and precision are generally contradictory; the more complex a system is, the less precisely we are able to compute its behavior from measurements of various inputs and conditions in the environment. Fuzzy logic models a system at a graininess more compatible with actual knowledge of the system and provides mechanisms for combining the "grains" in a meaningful way. Where precise, numerical output is needed, as in the case of control, methods for producing it have been successfully developed.

It has been an article of faith on the part of scientists and engineers for a long time that, if enough information were available about the state of a system at some time and a sufficiently powerful computer were available, the system's behavior at a later time could be predicted with certainty, apart from perhaps fluctuations arising from averaging over predictions to construct measurable quantities. However, recent studies of nonlinear systems—and much of nature is turning out to be nonlinear—have shown this faith to be unjustified. Often, the long-term behavior of a system is independent of how it starts. A limiting behavior in which there is no correlation between the system behavior at various times is called *chaos*. Most of the controlled systems are built not to behave this way, but do have nonlinear components that can cause problems for the control system in some situations. On the other hand, people often can control systems effectively using experience or by abstracting knowledge from computations that are not suitable for use in real-time control.

With such systems, Zadeh's methodology is attractive because it allows the imprecision (fuzziness) in complex systems to be addressed without distorting assumptions present in many other techniques. The mechanisms Zadeh introduced were the fuzzy set in which membership in a set could be partial and the extension principle that permits defining and manipulating relations between fuzzy variables, such as implication in the case of rules.

Membership in a set is often associated with properties describing the set. For example, the set of red fire trucks has the properties of vehicle type: truck, function type: firefighting, and color: red. Fuzzy sets remove contradictions at the boundaries between sets—multiple situations could be true to a degree at the same time. A red pickup truck or a green fire truck could have some membership value in the set of red fire trucks.

To illustrate this concretely, a mammal has hair and feeds its newborns with milk whereas a bird flies, lays eggs, and has feathers. These characteristics can define the sets *mammal* and *bird*. A duck-billed platypus has hair,

> **fuzzy logic: a mathematically based scheme for modeling and controlling complex systems by recognizing that complexity and precision are generally contradictory.**

> Heuristics, or rules of thumb, are an important component of human control. Rules similar to the old saying, "If it hurts when you lift your arm, don't lift your arm," work in many operator situations. In controlling a plant, operators handle vibrations and delays using similar heuristics. For delays, operators develop expectations about how long it takes for control changes to appear on sensor readings and are not bothered as long as the expectation is met approximately. These expectations might be of a form *about a minute* or *about eight seconds*. With training, they even learn to anticipate the final value from the early parts of change and can make adjustments while the sensor data is changing. In handling vibrations, operating at certain speeds can induce vibrations in equipment. An example is an imbalanced tire on an automobile. If a driver observes vibrations, she learns quickly to avoid those speeds. Although this is problem avoidance rather than positive control, it implicitly refers to an ability to meet goals, particularly fuzzy goals, by a variety of acceptable plan modifications. Being slightly late or arriving slightly early at a destination and waiting might not be optimal, but they are highly acceptable alternatives to damaging the car or the driver.

feeds its young milk, and lays eggs. It, therefore, is both a bird and a mammal, to some degree. Fuzzy logic would assign membership value between 0 and 1 to each possible set, as in

```
robin(mammal/0, bird/1)
platypus(mammal/0.8, bird/0.1)
gorilla(mammal/1, bird/0)
```

where the notation x/m(x) denotes the fuzzy set and its membership value. A value of zero means definitely no membership in the set and a value of one indicates definite set membership. The results are the same for fuzzy sets as for ordinary binary logic in which the evidence indicates a true/false answer, as in the cases of the robin and gorilla. Thus, fuzzy logic encompasses ordinary binary, or true/false, logic. Note also that the statement is that the individual has partial membership in both sets, not that the individual lies in the intersection of the sets, which is empty in this case.

The power of classification for reasoning is that we can obtain from a particular set of observations (called the current case or just the case) a partial match to one or more classes. If a case belongs to a class, the unobserved properties of the class can be reasonably ascribed to the case. If you observed an animal with feathers, it would be reasonable to assume that the animal lays eggs, even though you don't see it laying eggs. If you formed a working hypothesis that the animal is a bird, you could be sure about the robin, the platypus would bear some further watching, and the gorilla could be ignored.

While this might seem trivial and useless, there are at least two practical areas where the same reasoning is used to take action. First, fault diagnosis rests on classifying observations into classes that are the faults to be detected. Once a class is identified, there is usually a prescribed fix for the problem. The best fault-diagnosis system recognizes a fault before it has developed fully and allows corrections to be made before problems get unmanageable. Reasoning with partial information is de rigueur and fuzzy logic focuses consideration to potential options with some measure of priority of those options. Focusing can generate a plan for emphasizing a particular sensor subset to maximize discrimination between options.

current case: a particular set of observations made when trying to classify an object.

Second, control of a system can be accomplished by classifying sensor data into classes that lead to prescribed actions—such as heater on, heater off, cooler on, cooler off, and fan speed off, slow, or fast for a heating, ventilation, and air conditioning (HVAC) control system. Fuzzy control permits defining classes at sensor readings where the desired control action is well-defined and interpolating between the actions prescribed for different classes. This is the situation most of the time while the controller is operating because the sensor measurements match the conditions for the classes only partially. Later, I shall expand each of these to illustrate the ideas.

First, however, you need to understand how fuzzy logic processes information to establish the terms that will be used. Figure 11.1 shows the components of a fuzzy controller schematically. (A fuzzy classifier is similar but has a different defuzzification stage.) First, there needs to be a source of observations. Normally, this is a set of sensors. If I were building an HVAC controller, these might be room temperature, outside temperature, and humidity. These also can be fuzzy variables in that I can express them in fuzzy terms such as hot or cool. A fuzzy variable is a variable that can have fuzzy values. Next, I need to specify the ranges these sensors can reasonably produce. This is called a universe of discourse.

Figure 11.1. *The components of a fuzzy controller.*

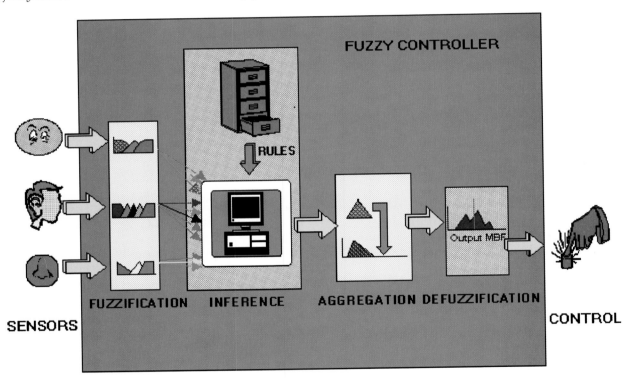

For example, I might choose the universe of discourse for the room temperature sensor to be 50 to 100 degrees Fahrenheit. I provide a name for each fuzzy variable associated with each universe of discourse. Often, the same name is used for the sensor and the fuzzy variable but, here, I shall change the name for clarity. I shall use RT to correspond to room temperature, OT to outside temperature, and HU to humidity. To RT, I ascribe fuzzy values of *cold, cool, moderate, warm,* and *hot*. To OT, I give fuzzy values of *cold, moderate,* and *hot*. Finally, to HU, I give fuzzy values *low, moderate,* and *high*. From these, I can create 5x3x3 = 45 classes. As I possess fewer controls, this might be too many classes. I can reduce the number of values for some fuzzy variables or simply disregard some of the possible classes when rules relating classes to actions are created.

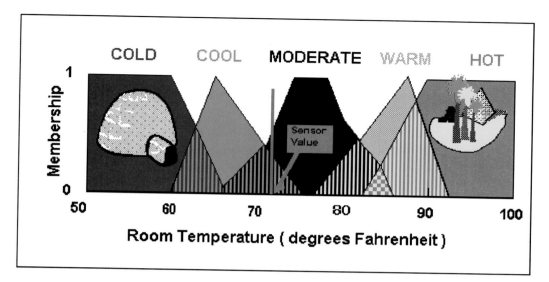

Figure 11.2.
Membership functions for the RT fuzzy values.

To relate sensor readings to the fuzzy values, the process of fuzzification, I introduce the membership function. Fuzzy logic does not constrain such a function, except that it be single-valued, but, for control applications, a simple triangle or trapezoid has proved effective. These can be specified by using 2, 3, or 4 parameters for symmetric triangles, general triangles, and trapezoids, respectively. Figure 11.2 illustrates some membership functions for the RT fuzzy values.

Rules are the next component of a fuzzy controller to be created. They are normally of the form, IF <conditions> THEN <actions>. I could use a form IF <conditions> THEN <class> and IF <class> THEN <actions> but, normally, there is a direct connection between classes and actions for applications. The conditions are often of the form, (RT is warm) AND (OT is cold) AND (HU is moderate), i.e., statements about a fuzzy variable having a particular fuzzy value joined by AND. Logical NOT is also possible as a modifier of any statement as in NOT (RT is hot). Logical OR is meaningful if the statements are about the same universe of discourse, as in ((RT is warm) OR (RT is hot)), but the use of OR to connect statements about different universes of discourse should be regarded as a shorthand way of writing several rules. Evaluation of the truth value of the left-hand-side conditions (LHS) is accomplished by evaluating for each sensor input its membership value in the designated fuzzy value and combining these by finding the minimum for statements connected by AND and maximum for statements on the same universe of discourse connected by OR. NOT is accommodated by subtracting the membership value from one.

Rules represent a mathematical relation between the left-hand-side (LHS) conditions or input and the right-hand-side (RHS) actions or output. The relation simplifies for sensor input of a single number. The output membership function is modified according to the truth value of the LHS. There are two popular procedures used. A procedure called max-min clips the output membership function at a level corresponding to the LHS truth value and max-dot multiplies each point of the output membership function by the LHS truth value. If the output membership functions are single values, the result is the same for both procedures. Figure 11.3 illustrates these processes.

All rules are evaluated in parallel and the outputs in the same universe of discourse must be aggregated to form the output fuzzy set. The membership value for (NOT A) is always 1-(membership value for A). These operations are valid whether the combination is for different output fuzzy values from several rules or the same output fuzzy value from rules considering different inputs.

Finally, for some applications such as control, a definite, or crisp, numerical value is required to drive further processing. Obtaining the value from the output fuzzy set is called

defuzzification: obtaining a numeric value from a fuzzy set.

defuzzification. A popular technique is the centroid method that provides the output value with equal areas above and below the point. Another is mean of maxima which provides a weighted average of the points with the same highest membership. Figure 11.4 illustrates these methods. The centroid method provides a smoother varying output with time as input conditions vary. The mean of maxima tends to provide step output as one set of conditions becomes dominant at a particular time. The defuzzification technique must be chosen by the designer and should be considered in the context of the problem. Defuzzification is a collapsing of the output fuzzy distribution into a single value and that value might distort the interpretation of the fuzzy output. For example, a collision avoidance controller for a robot might indicate clear paths on each side of an obstacle by large memberships at each end of the universe of discourse (each side of the obstacle), whereas using the centroid method instructs the robot to go through the obstacle.

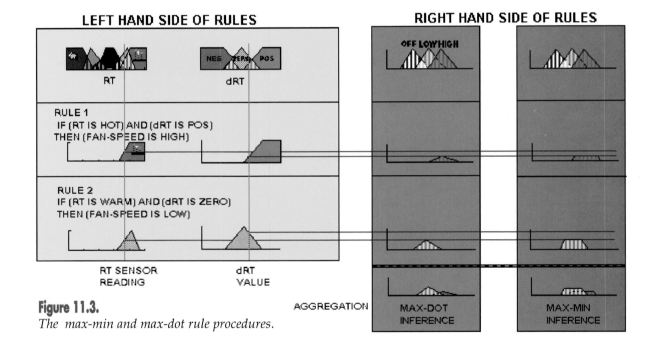

Figure 11.3.
The max-min and max-dot rule procedures.

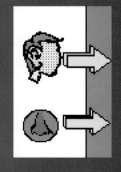

These are the basic elements of fuzzy logic applicable to control and decision support such as fault diagnosis. These techniques have been applied to consumer products such as washing machine control, air conditioner control, camera autofocus, video image stabilization, and vacuum cleaner suction control to industrial control such as ventilation fan control for highway tunnels, water purification plant control, cement kiln control, glass manufacturing control, subway train controls, and elevator scheduling. They have also been applied to space applications such as vehicle attitude control, tether mode detection and control, and temperature control for protein crystals growing on the Space Shuttle. These represent just a sampling of the applications; the number of fuzzy control applications is now very large and multinational in scope.

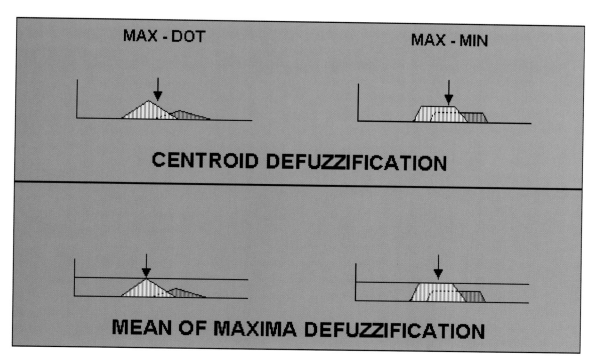

Figure 11.4
Two methods of defuzzification.

How are these elements combined to devise a controller or fault-detection system? You construct simple applications to show the process. In the case of controllers, I construct a simple controller for an HVAC system. The first task is to choose rules and names for fuzzy values for sensor and derived inputs. I try to use as few sensors as possible to keep cost and complexity low and add a new one only when I am convinced it is worth the cost increase. Clearly, here, the room temperature sensor is necessary. I hold the outside temperature and humidity as options only. I also can use past sensor readings to construct trends—direction and magnitude of the change of sensor values. I call this dRT for this case. The first rule is that if the RT is near the set point and the dRT is near zero, do no control, i.e., the fan is off, the heater is off, the cooler is off. I note at this point that I can base control on departure from a set point. I form such a fuzzy variable and call it Deviation. Deviation can have fuzzy values zero (ZO), small positive (PS), large positive (PB), small negative (NS), and large negative (NB). With five fuzzy sets, I can do nothing, run the fan slowly, and run the fan fast with heating and cooling, as appropriate. A motivation for including dRT is that when the room is being heated (cooled), dRT is positive (negative). But when the room is cooling (heating) due to outside temperatures, I want to run the fan slowly to circulate the air for overcoming convective layering of warm and cool air.

Some rules-of-thumb for fuzzy-controller design include using three–seven membership functions for fuzzy variables with positive and negative values, considering rules for all combinations of sensors although all might not be needed, and trying to use as few sensors as possible or grouping sensors into small groups to build subcontrollers whose outputs are combined by another controller. A rationale for the first heuristic is that there is a do-nothing, do-something, and do-something-big for positive and negative values if five membership functions are used. Three give rise to coarser grades of actions and seven are about at the limit of the designer's ability to keep options continuously in mind. The second heuristic prevents leaving out important combinations to assure reliable performance. Bart Kosko, a prominant advocate of fuzzy logic, advises that only about half of the possible rules are needed to achieve a satisfactory control. The final rule prevents trying to create a controller that is expensive and complex to develop and deploy. The number of possibilities is the product of the numbers of options for each input. For a four-input controller with five membership functions each, this number is 5x5x5x5 = 625, a hopelessly large number to handle effectively. If that is broken into two groups of two sensors, the number of possible rules is near 50, depending on the number needed to combine the subcontroller outputs.

For fuzzy diagnostic applications, the construction of membership functions and rules is similar to practices described for controller applications. However, defuzzification is not necessary and might even be detrimental. I aggregate the evidence and examine the support provided by the rules for each fault option. Normal operations and transitions between operating points and sensor failures need to be in the list of options. The system should call a fault only when the operation is not normal (m(normal) < 0.5 or some other threshold) and evidence supporting a particular fault is compelling (m(fault) > 0.5). Sensor faults can be overridden, as an interim measure, by instructing the fuzzifiers to produce membership values of one for all cases. This allows rules to consider the good sensors only and the fault detection to degrade slowly.

Fuzzy logic provides a basis for practical applications in control, decision support, image processing and compression, and robotic control. I have discussed mainly control but the fuzzy logic concepts are similar for other applications. Image processing and robotic control add the dimension of dynamic creation of membership functions by the fuzzy system. There is de-emphasis on using rules. An aspect of control I have not included is the ability to adjust the controller parameters and/or rules dynamically. The methods of doing so are at a research stage of development but offer bright prospects for future applications of this most fascinating technology.

CHAPTER 11

Fuzzy Logic

dr. jack aldridge

Dr. Jack Aldridge instructs courses in basic fuzzy logic and control applications and consults in the design and implementation of fuzzy control systems.

CHAPTER 12
Interactive Entertainment

mike morrison

In the near future, entertainment is about to get much more entertaining, thanks to modern technological advances. Even though interactive entertainment products such as coin-operated video games and computer entertainment software have been around for more than 15 years, recent advances in technology have opened up a way to merge the video and film industry with interactive entertainment. The result might prove to be an entirely new form of entertainment that will be as earth-shattering as the first television.

> **interactive entertainment:** games that allow the user to participate.

Major players in the commercial entertainment field such as MCA, Warner Brothers, and Paramount are moving into the consumer electronics field. At the same time, computer companies such as Apple, IBM, and Microsoft are investing in more and more interactive-oriented technologies such as multimedia and digital video. Likewise telephone companies such as GTE and AT&T, along with cable companies such as Bell Atlantic, Cablevision and Time Warner are pursuing ways to increase the bandwidth of existing telephone and cable TV wiring in order to provide interactive services.

The merger of these varied technologies is about to give birth to some exciting new products and services. First, lets take a brief look into the near future of interactive entertainment. Then let's step back to reality and look at current technology that is available today.

A View of the Future

In the near future, you might be able to choose which movies or television shows you want to view on demand. You might choose to watch an episode of "Quantum Leap." You are presented with a list of titles from all the previous episodes. After choosing one, the system prompts you for the type of view service you want. You can choose to watch the episode with "Maximum Advertising" (for a viewing discount of course), "Minimum Advertising" (for the standard viewing fee), or "No Advertising" (for an extra viewing fee).

Next you can choose the "Expanded Research" option. Choosing this option (for an additional cost) will allow you to pause the episode and view extra information relating to the time period the episode takes place in, the historical figures depicted, or hyperlinks to previous episodes for more information on a character you don't recognize.

Finally, the episode begins. In this episode Sam goes back in time to the Wild West. The show hasn't been on for very long however, before an advertising break starts. The local station node knows the basic theme of this episode is western, so it automatically runs an advertising spot for "Bob's Big Cowboy Hats." Finding a sudden interest in cowboy hats, you interrupt the advertisement and request a catalog. Back at your local station node, an automatic fax system faxes your name and address to "Bob's Big Cowboy Hats," along with a request for a catalog.

Because you've ordered a catalog, the advertisement aborts (as a reward for your voluntary submission to their mailing list) and you are returned to "Quantum Leap." As the episode continues, you notice a nice pair of western boots. You pause the episode and click on the actor's boots. A menu appears with the options to research the period dress of that time, or enter the Mall and go shopping. After choosing Mall, you see a list of stores that offer western clothing.

Once you choose a store, you are now in an online catalog. You select the footwear section, and after browsing through a number of boots, you find a pair similar to the ones in the episode. After verifying that they have your size in stock, you click the purchase option. After choosing your method of payment and delivery method, you are returned to the "Quantum Leap" episode. Now, however, you get awarded a free "No Advertising" view for the rest of the show because you made an online purchase.

As you can see, this new technology has benefits for the viewers, the providing service, and even the advertisers. It's a win, win, win situation.

CURRENT TECHNOLOGIES

Current interactive entertainment systems range from the office computer to the home video game system to portable game systems. Let's take a moment to look at the current state of the art in the interactive entertainment field and its acceptance by the adult community.

ACCEPTANCE OF INTERACTIVE ENTERTAINMENT

It's noteworthy that it has taken a long time for interactive entertainment such as video games to become accepted as valid forms of entertainment for adults. Take for instance, playing cards, which have been around for well over 600 years. It's perfectly acceptable to see young children, adults, or even older people in their 70s and 80s playing cards. This, however, is not the case for interactive entertainment, especially electronic devices such as video games. It is fairly uncommon to see adults and older people playing video games. This is about to change...

There is no question that the technology behind a video game is superior to playing cards. Playing cards in itself is very simple. You have 52 cards, divided into four major groups, and each group (or suit) is divided into 13 different levels. Yet with these few combinations, literally thousands of games have been invented over the years.

Now take the modern semiconductor that can contain billions of transistors on a single chip. Although semiconductors have been around only a few decades, people have wasted no time in using them to their full capabilities. One of the most lucrative uses for this technology was in video games. In 1976 simple, electronic pong games appeared in arcades throughout the country. At the same time a new market was found for the video game in the home. Atari introduced a home game system in 1979. The next three years (1980-1982) would see the video game industry jump from a $330 million-a-year industry to a $3 billion-a-year industry.

Still, the high technology involved with these devices was practically wasted on mindless shoot-em-up video games for kids. While the children blasted aliens in the living room, the adults played cards in the dining room. It was kind of a reverse psychology for technology.

This trend is changing as newer video games are starting to target more mature audiences. As you will see in the following sections, many video games are now being produced for the older, adult market.

HOME GAME SYSTEMS

Home game systems have progressed greatly over the past few years. Now the typical home game system features a 16-Bit graphics and sound engine, and a CD-ROM drive (see Figure 12.1).

Figure 12.1.
Here is the 16-Bit Sega-CD home game system with a CD-ROM drive.

Just last year a company called 3DO announced a home game system that will be released this fall (1993). Instead of simply making another game machine, 3DO is trying to bring the power of a high-speed graphics computer down to the price of a video game unit. Examine the following chart:

	Personal Computer	Game System	Television	3DO
Max Colors	16.7 Mil	256	2 Mil	16 Mil
Pixels per second	1 Mil	1 Mil	6 Mil	36-64 Mil

As you can see, the graphics display speed of the 3DO is order of magnitudes faster than current technologies. Added to the high-powered technology of the 3DO are the high-powered backers of the product. Time Warner, Electronic Arts, AT&T, and Matsushita (the parent company of Quasar, Technics, Panasonic, MCA-Universal, and many others) are partners in 3DO. Thus the makers of the 3DO plan on offering not just shoot-em-up style video games, but full length motion pictures that can be viewed directly from a CD-ROM. One of the first titles will be an interactive version of the MCA movie *Jurassic Park*.

Already companies such as Sega have released products such as the Sega-CD Virtual-VCR, which offers more than one hour of video on a single CD. They currently have titles such as *Prince*, which features music and footage of the popular rock star and his band. Also available is *March of Time* from *Time* magazine, which features old newsreels narrated by Orson Welles.

PORTABLE GAME SYSTEMS

Of all the new technologies responsible for the acceptance of video games for adults, the portable game unit is perhaps the greatest. While about 35 percent of dedicated game system players are adults, more than 46 percent of portable game players are adults. Likewise, portable game units appeal more toward adult females, with 44 percent of adult players being female as opposed to 29 percent for dedicated gaming systems.

Portable game units such as the Nintendo Game Boy (see Figure 12.2) started shipping in 1989. Following closely on its heels were the Atari Lynx, NEC Turbo Express, and the Sega Game Gear. All provide entertainment for the road with their cartridge-based games.

Figure 12.2.
The Nintendo Game Boy.

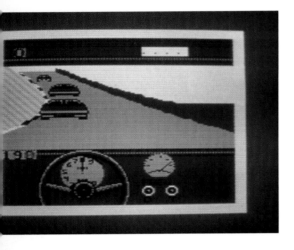

Figures 12.3-12.5.
Konami's Blades of Steel, Racing, and Ultra Golf, all available for the Nintendo Game Boy.

Anyone who has flown across the country in a plane knows that sometimes you just get tired of reading. If you are by yourself with no one to talk to, your options are pretty limited when it comes to keeping yourself entertained.

Game developers are likewise noting the trend toward older players and have responded with a variety of games geared toward the adult audience.

For example, Parker Brothers has released a cartridge for the Game Boy that allows you to play Monopoly with up to four players (see Figure 12.6). Likewise other companies offer various adult-oriented games, such as poker and chess, most of which require no hand-eye coordination whatsoever.

Newer game units such as Sega Game Gear offer high-resolution color displays, and even television tuners that allow you to watch TV if you get bored playing games.

Personal Computer Based

Games for personal computers have always taken full advantage of the computers' processing power. Due to this extra processing power, PC games have typically been more complex and involved. Whether it be a flight simulator that conforms to FAA's instrument flight training regulations, or a 3-D computer adventure with high resolution and computer rendered animation, personal computer games usually push the hardware to its limits.

The power of the computer has allowed a blend of entertainment and education. Such software designed both to educate and entertain is often called Edutainment software.

One such Edutainment package is The Miracle from Software Toolworks. The Miracle is a combination of hardware and software that you can add to your home PC. It comes with a full function MIDI (Musical Instrument Digital Interface) electronic keyboard, necessary cables, and software (see Figure 12.7).

The software comprises a year-long piano course that teaches you how to read music notation, play with two hands using chords and common rhythms, learn new pieces of music on your own, and perform with other musicians.

The software can communicate to the keyboard and detect which keys you play, when you press them, how hard you press them, and when you release them. This information gets passed on to the computer which in turn uses artificial intelligence routines to analyze the way you play. After determining the most significant errors, the software recommends and sometimes creates specific exercises to improve your playing.

Interactive Entertainment

> The beauty of the portable units is that you can play them anywhere, at the pool while getting your tan, at the amusement park while waiting in line, at the garage while waiting for your oil change, practically any time or location away from home when you have to wait on something.

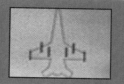

In the realm of pure entertainment, many game developers are taking advantage of the latest technology such as 3-D computer animation and digitized live video.

Virgin Games has recently released an interactive adventure game called The Seventh Guest. It comes on two CD-ROMs, which contain a total of 1.2 Gigabytes of data (equivalent to 360,000 pages of text). The reason the game requires so much storage is that it offers high resolution 3-D computer animated sequences that allow you to explore an entire house.

For instance, you might find yourself in the foyer with a beautiful view of a large marble staircase. Clicking on the staircase will smoothly walk you over to the staircase and glide you up to the top. From there you can walk down the hallway, and explore various rooms in the house (see Figure 12.4).

Other companies such as ICOM Simulations and Infocom are combining live video footage with computer games. Infocom for example is about to release *Return to Zork*, an interactive adventure game that combines 3-D computer animation and live video. Professional actors, screenwriters, and directors were used for the production along with make-up, costuming, prop, and camera crews. Scenery experts also were hired for the production. In many scenes, the actors were filmed in front of a blue screen then later a 3-D computer generated background was added (see Figures 12.9 and 12.10).

edutainment software: computer games or software that not only entertain, but are also educational.

Figure 12.6.
Monopoly for the Nintendo Game Boy.

Figure 12.7.
The Miracle from Software Toolworks.

On the Cutting Edge of Technology

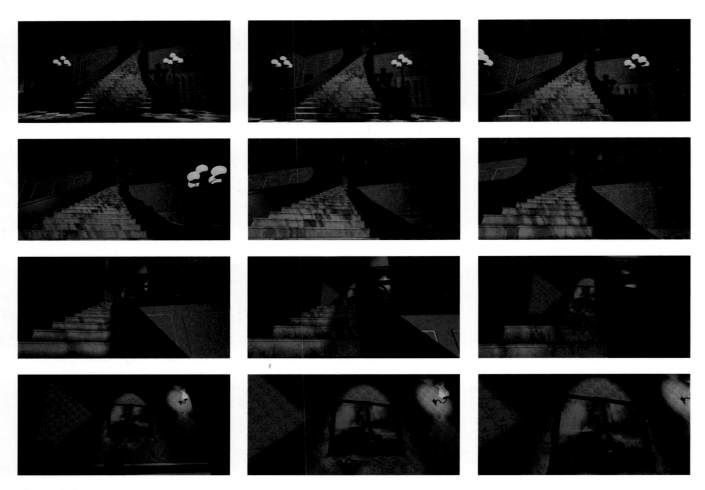

Figure 12.8. *This is a 12-frame sequence showing the smooth 3-D computer animation used in The Seventh Guest.*

 ICOM Simulations prefers to film the actors within real sets exclusively such as in their *Sherlock Holmes: Consulting Detective* series. In Sherlock Holmes, the player gets to guide and direct Sherlock on a variety of interviews with possible suspects to solve a given case. The interviews are live video clips that play directly from the CD-ROM. Once you have determined who committed the crime and how it was committed, you appear before the judge to plead your case. If you are correct, the computer grades you on how quickly you solved the mystery (the number of interviews it took).

flight simulation:

a program or machine that provides the illusion of flying.

One way the game developers are advancing PC-based games is by using non-linear storylines. A good example of this is the new *Strike Commander* game from Origin Systems. Not only does it use some of the latest technology for 3-D flight simulation (see Figure 12.11 and 12.12), but it also provides a storyline that changes based on your performance. If you fly the missions well, you make money and the plot takes a more positive turn. If you do not accomplish all the objectives on the mission, the plot takes more of a negative turn. Thus, the storyline and plot may be completely different for any two players of *Strike Commander*.

Interactive Entertainment

Figure 12.9.
Here, actress Robyn Lively, from the TV series Doogie Howser, is filmed in front of a blue screen.

Figure 12.10. *Afterward, the actress is overlaid onto a 3-D computer generated scene.*

On the Cutting Edge of Technology

Figures 12.11 and 12.12.
Two scenes from the game Strike Commander from Origin Systems.

VIRTUAL REALITY BASED GAMES

Virtual reality has become an overused term in the field of computers, but it still seems to be a driving force in the entertainment field. A number of companies are currently working on new entertainment products that seem to be a cross between an arcade game and a super computer.

One such company is Magic Edge in Mountain View, California. Magic Edge is currently working on a Hornet-1 flight simulator very similar to the ones that the military uses. A high-speed graphics workstation provides high-resolution 3-D views inside the cockpit on a 40" screen (see Figure 12.13). The cockpit itself is mounted on a platform that allows it to move 60 degrees in any direction (see Figure 12.14). Multiple simulators can be networked together to allow players to dogfight each other. Rides in the Hornet-1 simulator probably will cost about $25 and last for about 15 minutes.

Interactive Entertainment

Figure 12.13.
The view from inside the Hornet-1 simulator.

As you can see, video games are no longer for kids only. Finally adults can start benefiting from some of our high technology that is usually reserved for business. From hand-held portable video games to million-dollar virtual reality systems, interactive entertainment is here to stay.

Figure 12.14.
The cockpit of the Hornet-1 simulator moves up to 60 degrees in any direction.

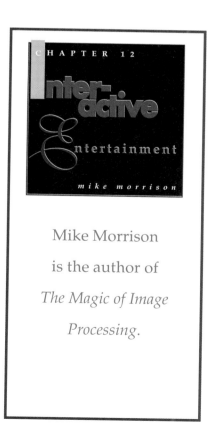

Mike Morrison is the author of *The Magic of Image Processing.*

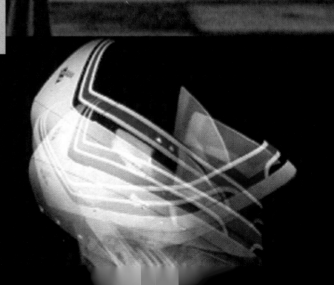

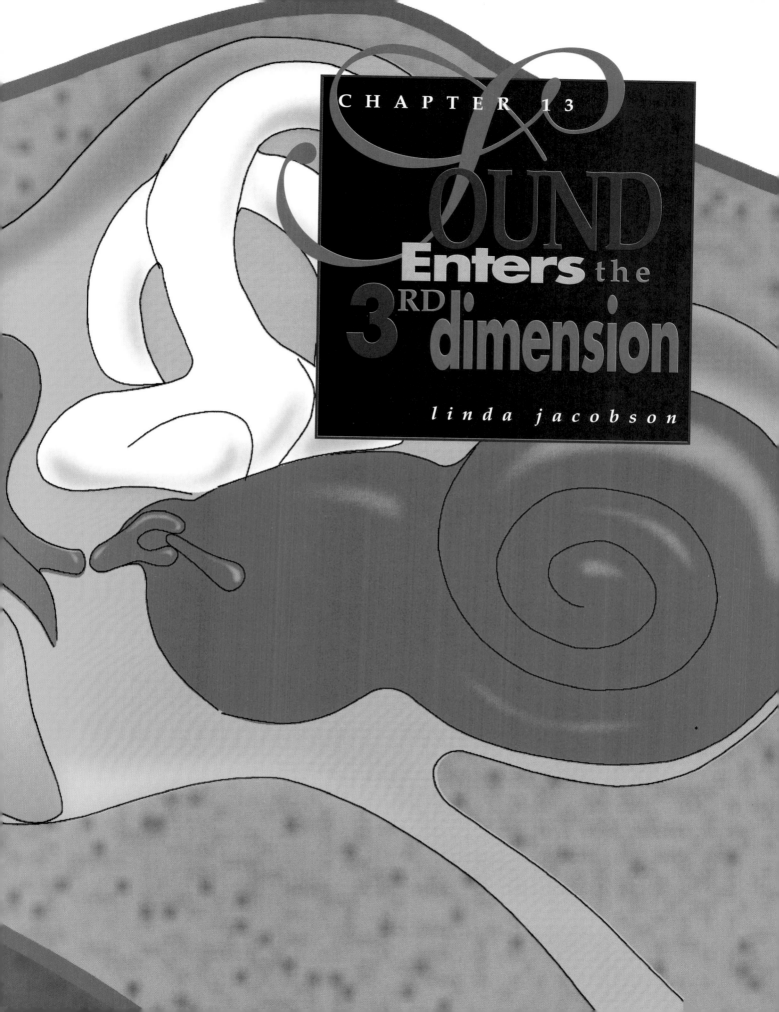

CHAPTER 13

Sound Enters the 3RD dimension

linda jacobson

Shut your eyes and listen. Hear that buzzing bee? As it slowly circles your head, the roar of a jet soars above you. Both sounds fade into the distance. Suddenly, just inches from the side of your face, a soda can pops open and the liquid sizzles into a cup. You can practically feel the sticky mist on your cheek. Hey, what's that music? It's a brass band, its faint melody growing louder as the musicians approach you. Within seconds the ensemble seems to surround you, the saxophones honking on your right, snare drums ra-ta-tatting on your left, trumpets blaring in front of you. You raise your arms to conduct the music—watch it, don't get tangled in the headphone cord!

When you remove the headphones, you marvel at the way it all sounded so three-dimensional. You just experienced 3-D sound. But is it cutting-edge wizardry...or manipulative sonic trickery?

When you listen to a 3-D sound recording on a normal stereo system, you hear the sounds as if they were coming from different places all around the room. Above you, below you, beside you, before you—even from some point far off in the distance. Seldom do the sounds seem to emanate directly from the speakers. When you hear the same recording on headphones, the effect is even more uncanny. Sometimes it is impossible to distinguish the recorded sounds from "real" ones. That's what 3-D sound is all about (see Figure 13.1).

Also known as spatially enhanced stereo, three-dimensional sound results from an audio manipulation process that lets recording artists enlarge their musical canvas and paint a more realistic sonic picture. The process lets listeners enjoy the results on garden-variety, two-speaker stereo systems, without adding loudspeakers, special headphones, or fancy sound decoders.

The new 3-D recording and production systems—which started appearing on the market in the early 1990s—are said to do everything from expand the stereo imagery to make sounds seem to jump from the speakers and swirl around the room. Even if they don't make the sound swoop about your head, they at least enhance the perceived presence of music played on ordinary stereo systems. Using 3-D sound technology, the recording artist can "place" sounds in three-dimensional space, controlling their direction, distance, and depth, the motion of the sound source, and the aural image's overall size, stability, consistency, and positioning.

When people say stereo, they're usually referring to the kind of dual-speaker systems introduced in the 1950s, six decades after Thomas Edison invented the phonograph record. The word stereo comes from the Greek stereos, which means "hard," "firm," or "solid." When combined with other words—creating stereophonic, for example, or stereoscope—it alludes to the notion of three-dimensionality. Since the late 19th century, music lovers have tried to boost the dimensionality, fidelity, and clarity of recorded sound. Quality dramatically improved with the development of "high fidelity" phonograph records in the 1940s. Within ten years stereophonic records were all

Figure 13.1.

The difference between listening to 3-D sound versus regular stereo sound.

the rage; two or more microphones were used for recording, and two speakers were used for playback. Ever since, even when digital compact discs blasted onto the music scene, stereo has been the absolute sound standard.

A stereo system emits a different mix of sounds from each of its two loudspeakers. This helps clarify the sounds of various musical elements and gives the impression of the musicians and singers performing in different positions in front of you. In other words, stereo conveys a sense of space.

In the studio, when engineers mix various instrument recordings to produce for stereo, they spread the musical elements along an imaginary flat line. That line extends between the two speakers; one at the far left, the other at the far right. How do the engineers spread sounds within this horizontal sound field? On the audio mixing console, there is an adjustment labelled "pan" (short for "panoramic"). While mixing the sounds, the engineer adjusts the pan to place each instrument in the sound field—toward the left, right, or center. To hear the effects of panning, listen to a stereo record, CD, or cassette as you twist the balance control all the way to the left, then all the way to the right. You'll hear how different sounds were panned in each direction, or (especially in the case of pop music vocals) kept in the center of the sound field.

The new 3-D sound technology blurs the flat line of that sound field. During the recording or mixing process, engineers use a 3-D sound processor to spread the sound above, below, and beyond the edges of the speakers. The 3-D systems are used to place, or localize, sounds in the "up," "down," "front," or "back" positions. Some systems re-create the original recording environment, providing an acoustic photograph of live music and sounds as they filled a room or concert hall. (*Omnidirectional*, then, might be more appropriate than three-dimensional, but "OD" just doesn't sound as appealing!)

All the 3-D sound systems work in different ways but share the same goal: to trick your brain. They fool you into perceiving that the sound is coming from somewhere other than the speakers or headphones. The trick works because of recent advances in computer technology, signal processing, and psychoacoustic research. Signal processing is a technique used in the studio to change the characteristics (such as pitch) of an audio signal. (Audio is the electronic representation of sound.) Psychoacoustics is the branch of psychophysics that explores how we perceive and interpret sound. It's related to acoustics, the branch of physics concerned with the properties, production, transmission, and effects of sound. When we talk about a sound's "warmth," "clarity," or "punch," we're talking subjectively. That's psychoacoustics. When we talk about a sound's intensity or echo, we're talking objectively. That's acoustics.

To understand three-dimensional sound, we need to know how we hear. Ever wonder about those smooth shells of flesh on the sides of your head? Just as we possess stereoscopic vision, we also have stereo hearing. Our ears are perfectly designed to capture sound. Sound is created from regular vibrations of air; these vibrations are called soundwaves, or cycles of alterations in air pressure. The ancient Greeks originated the concept that sound is transmitted through air, reaching the ear to produce the sensation of hearing.

The shell-like shape of your outer ear—the *pinna*—suits it to the job of collecting soundwaves from different directions and directing those sounds down varying paths through the ear canal (see Figure 13.2). The pinna actually modifies the soundwaves, helping the brain determine where the sound came from: front, back, up, or down. The ear canal acts as a resonator. Like a pipe organ in a cathedral, it echoes and enriches the soundwaves. When soundwaves reach the canal's end, they hit the eardrum, a thin, vibrating membrane. This causes varying pressure on the eardrum, which transmits the pressure to the middle ear. (Depending on the sound's volume, the eardrum's vibrating movement varies from 10 to 40 millionths of an inch.) Inside the middle ear are three tiny bones, or *ossicles*, called the hammer, anvil, and stirrup. They pass the

Figure 13.2.
The ear and its inner workings.

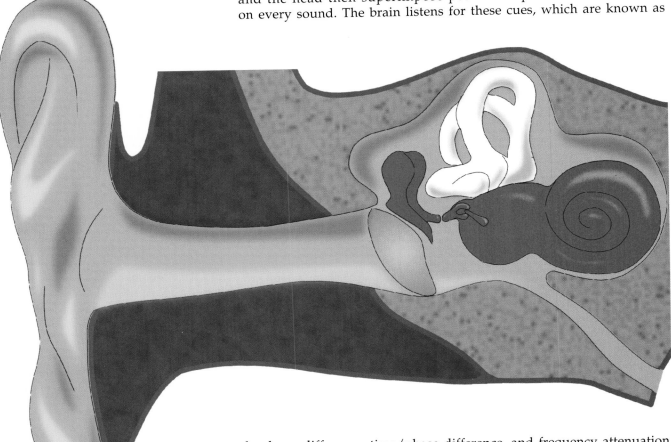

vibrations of the eardrum into the liquid-filled inner ear, or *cochlea*. Inside the cochlea, the basilar membrane changes shape in response to the liquid pressure. Its movement excites the hair-like nerve fibers in the inner ear, causing impulses to "fire" at the brain's acoustic cortex.

Both ears capture soundwaves that come from all directions. The pinnae and the head then superimpose position-dependent sound "cues" on every sound. The brain listens for these cues, which are known as loudness difference, time/phase difference, and frequency attenuation. In a recording, *loudness difference* occurs when studio engineers pan sounds to create a conventional stereo image. (They make the sound's volume louder in the right speaker or louder in the left.) *Time/phase difference* refers to the fact that soundwaves don't strike both ears at precisely the same time. *Frequency* refers to the number of times each second that the soundwave hits our ears; *attenuation* means reduction. *Frequency attenuation* is the process by which the ear's shape slightly changes certain frequencies of the soundwave, depending on the direction the wave comes from. Using these three cues, the brain can determine—*localize*—the exact location of the sound's source.

Let's say you're about to cross a two-way street, when suddenly a horn beeps. Without looking for the horn's source, you retreat and wait for the car to pass. The horn's honk reached your left ear before it reached your right ear, which helped you localize the sound and realize that the car was headed at you from the left. Sound travels at around 1,100 feet per second; sounds coming from the left reach the left ear almost a millisecond before they hit the right ear. When the sound does reach the right ear, it seems quieter because of the head's "shadowing" effect. This relates to the time/phase difference mentioned earlier. It is the difference in wave arrival time, or phase difference, that—when combined with loudness difference—cue your brain about where the sound originates.

> **To get an idea of how well it all works, try this experiment. Close your eyes and ask a friend to snap her fingers in front of your face and keep snapping while she moves her hand slowly to a point above your head. If you can tell where each finger-snap comes from, your pinnae are functioning just fine.**

Another effect also shows how much we rely on our pinnae. Called the "cocktail party effect," it enables us to enjoy an intimate discussion while at a noisy gathering, subconsciously filtering out the whoops, hollers, and guffaws around us so we can concentrate on the whispering at hand.

Many other effects result from the shape and placement of our ears, reflections of sound off our shoulders, the distance between our ears, shadowing effects of the head on opposite-side soundwaves, and resonances in the outer ear and ear canal. Taken all together, these factors create the *head-related transfer function*, the scientific term for what happens when our ears scoop up sounds and our brains process those sounds.

Enough about our two ears. Let's return to three-dimensional sounds. For over a century, people interested in sound have been trying to re-create "3-D" sound. In the 1880s they started making binaural recordings. Binaural refers to the use of two ears; a binaural recording is a stereo recording made with two microphones that are positioned to emulate the sensitivity and spacing of two ears on an average human head. Sometimes the microphones are even placed on or in the recording engineer's own ears. Preferably, the process involves an artificial or "dummy" head, an odd-looking piece of equipment modeled after a human head outfitted with a tiny microphone in each ear (see Figure 13.3). Each microphone feeds a separate audio track on the recording tape or other medium. At the listening end, these tracks are fed to separate sides of headphones; sounds captured by the original left microphone are sent to the listener's left ear, whereas sounds captured by the right microphone go to the right ear.

binaural recording: a stereo recording made with two microphones that are positioned to emulate the sensitivity and spacing of two ears on the average human head.

Music lovers have never particularly cared for the results because you must wear headphones to enjoy the effect. (For that matter, quadraphonic or four-channel sound didn't catch on in the 1970s because few people wanted to buy extra speakers and amplifiers for their stereo systems). Nonetheless, musicians experimented with the process. *Lou Reed: The Bells*, for instance, released in 1979, was recorded in West Germany using a dummy recording head. Other 1970s and 1980s

Figure 13.3.
On the right, a photo of "Fritz," the Neumann K100 Dummy Head, and NASA Ames Research Center 3-D sound scientist Durand Begault. Above, the back of Fritz's head. (Photos courtesy of Linda Jacobsen.)

binaural recordings were released by Steve Winwood ("Higher Love" in *Back in the High Life*), Stevie Wonder (several tracks on *Journey Through the Secret Life of Plants*), Pink Floyd (tunes on *The Final Cut* and several other albums), Kiss, and Quincy Jones. In 1985 Michael Jackson's *Bad* used all manner of binaural recording, resulting in stunning 3-D effects that required headphones to hear. (For a great example, grab some headphones and the album and listen to the spoken introduction of "I Just Can't Stop Loving You.")

Binaural recording systems aim to repro-duce the ambience of the original sound environment. Different sound equipment manufacturers have concocted their own recipes for serving 3-D over two speakers. Some systems use dummy heads; others, through the miracles of modern computer science, electronically simulate the binaural recording technique by applying signal processing in the digital domain of a computer. All attempt artificially to duplicate the natural triggers that help the brain localize sound.

interaural crosstalk: when sounds from the left speaker reach a listener's right ear and sounds from the right speaker reach the left ear, spearing the sonic imagery of a three-dimensional recording.

Binaural recording works best on headphones because it re-creates the aforementioned head-related transfer function right at the listener's ears. That's why an acoustic problem crops up when you listen to a binaural recording over two loudspeakers. The problem is called interaural crosstalk: Sounds from the left speaker reach your right ear while sounds from the right speaker reach your left ear. This smears the sonic imagery. On top of that, the room, even the furniture in it, affects the way you hear what issues from the speakers.

However, thanks to computer science, the new 3-D systems can mimic our auditory system (ear + brain) by supplementing the binaural process with *transaural processing*. They re-create spatialized sounds for listening over two loudspeakers by preparing specially structured correction signals and applying them to the speakers. Phase information is manipulated to effectively eliminate interaural crosstalk. (*Phase* describes the relative position of two soundwaves with respect to one another.) Some recording studio wizards use transaural processing independently to process binaural recordings, so they sound as realistic when played back via loudspeakers as they do on headphones.

To summarize, the new 3-D sound systems duplicate the localization cues provided by our bodies, emulate the interaural time delays and filtering effects that occur when we hear sound, and electronically re-create those effects in real time. These systems provide knobs, joysticks, or similar control devices (sometimes in software form, represented on a computer screen) to let the recording engineer place the sounds in the "up," "down," "front," and "back" positions. With these controls the engineer can even make a sound slowly or rapidly "move" from one direction to another (see Figure 13.4).

Research into 3-D sound has intensified in the past decade. A handful of manufacturers now market commercial systems to recording artists and studios, who use the new tools to produce 3-D recordings that will be sold in record stores. Trail-blazers in 3-D sound research include Roland Corporation, Head Acoustics of Germany, Sennheiser, Neumann, Desper Products, B&K, Kemar, Audio Cybernetics, Crystal River Engineering, and Focal Point.

All these new systems encode the process directly onto the recording medium, eliminating the need for the listener to buy a special processing device. Most systems combine loudness difference with either phase difference or frequency attenuation to reproduce the aural localization cues. Unfortunately, using only two of the three cues results in a less complete sonic "photograph." When a system combines loudness and phase differences, requiring two speakers to provide phase cues for a single sound, the "sweet spot phenomenon" occurs: the listener's head must be situated exactly between the two speakers and within a certain distance from them to accurately decode the cues. Move a few inches in any direction and the effect diminishes.

"3-D CDs" sometimes include instructions explaining how far from the speakers you should sit and how to arrange your stereo system to achieve the best results (see Figure 13.5). These results are obtained when the room is fairly "dead" (acoustically absorbent), especially near the speakers. If you place the speakers too close to a wall, sound reflected off the wall cancels out the phase cues and the 3-D effect diminishes greatly. You also lose

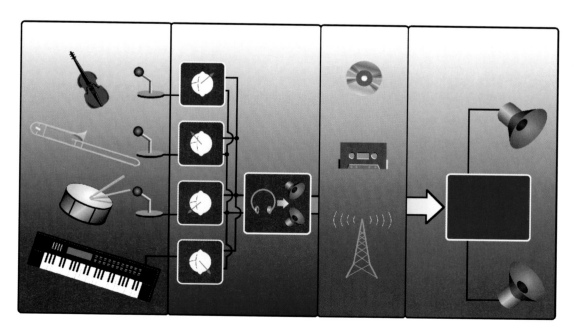

Figure 13.4.
A diagram showing how sounds are fed through 3-D sound system and output on two speakers.

the full effect when you listen to a 3-D recording inside an automobile (unless you sit in the middle of the back seat between the two rear speakers). And the effect doesn't work well over radio because stereo signals automatically blend into mono when your radio is not within the radio station's transmitter range (10–15 miles).

Systems that combine loudness difference and frequency attenuation require changes in the actual sonic structure of the recorded sound. Not many musicians and record producers willingly accept alterations of the sound for the sake of a 3-D effect. Besides, 3-D sound puts a new twist on music arranging, production, and recording, adding a whole new element to the production process.

Finally, this equipment isn't cheap. The typical 3-D sound processing system costs tens of thousands of dollars. Some recording producers, studios, and record companies consider 3-D sound an expensive gimmick.

This new sound technology affects more than music. Its potential certainly intrigues movie producers. (However, don't confuse 3-D sound with Dolby or THX sound, which involves many more than two speakers.) On stereo TV broadcasts and in video games and theme park rides, moving audio elements enhance the visual images. In acoustic research, sophisticated dummy-head recordings help researchers evaluate the performance of communications devices such as telephones, headphones, hearing aids, hearing protectors, microphones, and even motor

Figure 13.5.
Suggested optimum speaker placement and listener position for 3-D sound enjoyment.

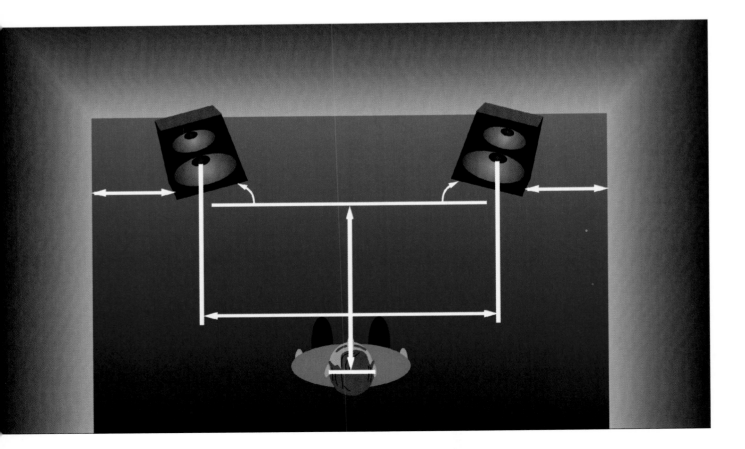

Sound Encounters the Third Dimension 145

These limitations haven't stopped artists who want to perch on the cutting edge. Among the first to use the new 3-D sound systems were Michael Jackson (on the album *Dangerous*), Madonna (on *Immaculate Collection*), Sting (on *Soul Cages*), and Luther Vandross (on *Power of Love*). Jackson used a 3D system to dramatize sound effects and spoken vocals. New age artist Suzanne Ciani used one on her 1991 album *Hotel Luna* to place the sounds of waves, wind, and raindrops.

vehicle noise measurement and analysis. Since 1981, German automotive firms have used Head Acoustics' dummy recording head to analyze interior and exterior car noise, judging the subjective annoyance of noises associated with each design. Their work has helped produce quieter automobiles.

As 3-D sound imparts information, facilitating *situational awareness*, it's also used in aerospace and aviation centers and virtual reality systems. Scientists at top research facilities tend to focus on headphone applications of 3-D sound, rather than speakers. At NASA-Ames Research Center in Moffett Field, California, researchers have been working since the 1980s on headphone-based 3-D sound systems that help people manage information in situations where spatial awareness is important, especially when visual information is limited or absent or to lessen the visual overload of complex instrumentation—such as in air traffic control displays in towers and cockpits and the monitoring of robotic activities in hazardous situations.

One such system was developed at NASA Ames and now is marketed commercially by a Northern California company called Crystal River Engineering. The *Convolvotron* is a pair of circuit cards designed to install in a standard IBM-compatible PC. (Convolution refers to the filtering of an input signal by head-related transfer functions.) The "Convolvotron" presents 3-D audio signals over headphones to match the head motion of the listener and/or the motion of the sound sources. As you move your head, the perceived location of the sound source remains constant: when you turn to the left, sound that you first perceived to come from in front of you then seems to come from your right. The system is very expensive, priced for well-endowed research labs and similar institutions.

In early 1993, however, Crystal River introduced the $1,800 "Beachtron" for personal computers, consisting of a circuit board containing a sound synthesizer, sound recording/playback functions, and a software library. The Beachtron lets you spatialize two separate sound sources for binaural presentation over conventional headphones. You can combine up to eight Beachtrons for 16 simultaneous sound sources, and you can record sounds into the PC directly off CD or tape, or live, or create sounds in the computer using the synthesizer software. Like the Convolvotron, the Beachtron works with other types of software programs, such as virtual world-building systems for virtual reality applications.

There are other relatively low-cost, 3-D sound circuit cards you can pop into your personal computer. One such product is the Focal Point system from Focal Point 3-D Audio of Niagara Falls, New York. It lets IBM PC and Apple Macintosh owners explore CD-quality, 3-D sound for under $2,000. The system processes mono sounds so they seem to come from three dimensions, with direction, elevation, and distance cues built into the audio signal. The sound plays back on any two-speaker stereo system or through headphones. Focal Point even includes mouse- and head-tracking support for virtual reality applications.

Meanwhile, other small companies around the country are developing and marketing similar affordable products that let computer-savvy folks experiment with 3-D sound. In the future, experts predict that recording artists and studios will be able to buy 3-D sound chips that pop directly into audio mixing consoles and musical keyboards. Others foresee the day when groups of people will be able to walk around in a room equipped with two speakers, simultaneously sharing an audio illusion and experiencing the full sound field of an original recording. And, one day, we may be able to create our own spatial sound mixes at home by manipulating a few controls on our stereo receivers or home entertainment control systems.

As the goal of using 3-D sound is to impart a sense of sonic depth and presence, and increase intelligibility and accuracy, by improving audio quality and finding new ways to manipulate sounds we enhance the overall listening experience and affect the listener's perception of reality.

In a way, 3-D sound technologists pay homage to the likes of Edgar Bergen and Charlie McCarthy, Jerry Mahoney and Knucklehead Jones, Waylon Flowers and Madame, the stars of ventriloquism. After all, when we consider the definition of ventriloquism—"the art or practice of speaking so that the voice seems to come from some source other than the speaker"—we realize that 3-D sound is a form of electronic ventriloquism. The analogy is even more apt when we talk about 3-D systems that use a dummy!

Whether 3-D sound truly transcends stereo, alters our perception of reality, or just serves as a recording studio tool to manipulate sounds and enhance the listening experience, it marks the beginning of an exciting era in sound technology.

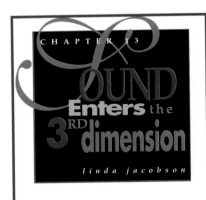

Linda Jacobson is the editor and co-author of *Cyberarts*.

CHAPTER 14

CHAOS AND COMPLEXITY

bill roetzheim

The world is not a simple place. Your body consists of trillions of cells all interacting to keep you alive and healthy. Your mind consists of billions of neurons interacting to allow you to think and remember. The stock market consists of millions of people like you, all interacting to create a market with behavior that virtually every analyst describes as if it were a living entity.

Traditional science focuses on understanding the individual pieces of a problem. How does a cell work? How does a neuron work? How does an individual investor behave? Tremendous strides have been made in answering these questions. The next logical step was to take knowledge about the individual components and use that knowledge to understand the behavior of groups of components. That didn't work. No matter how thoroughly scientists analyze each of the types of cells in your body, they get no closer to understanding how these cells interact. For example, what makes one cell decide to be a liver cell while a nearby cell decides to become part of a blood vessel? No matter how thoroughly scientists analyze each of the neurons in your brain, they get no closer to understanding how human thought works. For example, how does memory get stored? No matter how thoroughly economists analyze individual investor behavior, they get no closer to understanding the behavior of a large market such as the stock market. For example, what causes the market to suddenly and unexpectedly fall 500 points? The answers to these and a myriad of other real-world questions might be found in a new field of science called **complexity theory**.

A Working Definition of Complexity

complexity science: the study of emergent behavior exhibited by interacting systems operating at the threshold of stability and chaos.

Complexity is so new that authors of existing books haven't really tried to define it. If they worry about it at all, they use analogies and examples to convey the general concept. Webster's defines complexity as "...anything complex or intricate..." and defines complex as "...a group of related ideas, activities, things, etc. that form, or are viewed as forming, a single whole." Although my definition of complexity as it applies to complexity science is similar, it is somewhat more precise.

To understand this definition, and hence the science of complexity, you need to understand each of the three components of this definition.

The Study of Emergent Behavior...

Seemingly complicated behavior can be quite predictable in actuality. The automatons at Walt Disney World move, speak, and generally behave in an amazingly lifelike way, yet each of their movements and gestures was preordained at the time their program was written. They respond in a preprogrammed way to the environment and have no ability to adapt to changes. If the individual who programmed them watched them for hours on end, he or she would never observe anything that was surprising or unexpected.

Complex systems exhibit complicated behavior that is not predictable. A programmer that develops a complex model on the computer will frequently observe behavior that is surprising or unexpected. The behavior was not programmed in from the beginning, it emerged as the program operated. I term this unexpected, nonprogrammed behavior emergent behavior. The world of complexity is filled with examples of emergent behavior. You will see some simple examples later in this article when I look at the program Tapestry.

...EXHIBITED BY INTERACTING SYSTEMS...

The behavior of individual things is best studied using traditional science, and their behavior will seldom be surprising. An individual cell behaves in a predictable fashion, as does an individual neuron and, to a large extent, an individual investor. It is only when these individual components interact with each other, and influence each other through these interactions, that the system as a whole can begin to exhibit complex, emergent behavior. The science of complexity studies, by definition, the behavior of the system, as opposed to the individual components, as the individual components interact.

...OPERATING AT THE THRESHOLD OF STABILITY AND CHAOS.

The interaction between components can be structured to force the system into a static or repeating state. Perhaps the interacting pieces all eventually turn red. Perhaps they cycle between red and blue. As you slightly perturb the system (change it), it eventually returns to the same stable state. As you perturb the system in a major way, it eventually returns to the same or another stable state. Stability is interesting, but it is not complex. For living systems, stability is also not adaptable to changes. In complex systems that exhibit lifelike, emergent behavior, stability is death.

The interaction between components also can be so intense and varying that the individual components exhibit quasi-random behavior. The system as a whole consists of a myriad of components interacting in a seemingly random fashion with no patterns, no memory, no emergent behavior. These systems are said to be chaotic. In living systems chaos prevents the organism from retaining traits that are good. In complex systems, chaos is often pretty but largely useless.

A single interacting system will be stable with certain parameters, and chaotic with others. As the parameters are adjusted from those that result in stable behavior to those that result in chaotic behavior, there is a transition in behavior for the system. At the exact values where this phase transition is occurring, the system begins to exhibit the emergent behavior that I characterize as complex. It is chaotic enough to adapt and seek out new behaviors, but stable enough to hold onto those that work.

All living systems operate continuously in this narrow phase transition region between stability and chaos, this region of complexity. They don't exist there because they chose to, but because the interacting forces operating upon them force them to this region. Stable interactions

---

emergent behavior: unexpected or surprising results from a complex system that wasn't programmed into the system, but instead emerged after the program began.

system: a collection of individual components that interact with and influence each other.

chaotic: a system of components interacting in a seemingly random fashion with no patterns, no memory, and no emergent behavior.

are forced to become more chaotic. Chaotic interactions are forced to become more stable. The system finds itself pressed between two conflicting demands, driven inexorably to balance in the narrow, but optimal, region between stability and complexity.

In the remainder of this chapter, I explore stability, chaos, and the complex transition region in more detail.

Stability

A stable system is one that both exhibits repetitive behavior and that tends to return to this behavior when artificially forced to a different behavior. For example, an ordinary pendulum is a stable system because it tends to hang straight down forever, and when forced to swing using a push by your hand it will eventually return to its stable, hanging-down position. The hanging-straight-down position for a pendulum is called a point attractor because the pendulum is drawn to a specific point directly beneath the weight.

I'll demonstrate stability and attractors using a program called Attractor. The Attractor window, shown in Figure 14.1, is divided into four sub-windows. On the left, there is a plot for the X, Y, and Z parameters as they are calculated. On the right, you can see a window where the X and Y parameters are plotted in two dimensions. This figure shows an example of a point attractor. From a wide range of initial conditions, the equation results are drawn inexorably to a single point. For Figure 14.2, the equations were altered slightly to generate a circular attractor. From a wide range of initial conditions, the equation results are drawn into this circular orbit.

Not all attractors are as simple as points or ellipses. Figures 14.3 and 14.4 show two examples of a class of attractors called strange attractors because they do not correspond to a simple geographic shape. Figure 14.3 is an example of a Lorenz attractor, and Figure 14.4 is an example of a Julian attractor. It is important to note that these attractors are clearly stable in spite of their strange appearance.

Chaos

You might be wondering what exactly I mean when I say a system has become chaotic. Is it random? No, in fact, far from it. If I use X to represent the equation's result, the formula for predicting future values of X is still just as simple. Given a value for X_n, I easily can plug it into the equation and determine the value for X_{n+1}. So I can accurately calculate the value for X at a given future step, right? Wrong! The problem is simple but profound. For chaotic systems, very minor variations in the initial parameters have an ever-increasing ripple effect on the final results until eventually, the final results are as dependent on the initial variation in measurement as on anything else. Increasingly accurate initial measurements will allow you to make forecasts more and more iterations into the future, but you always will be eventually overwhelmed by that initial measurement error, no matter how small.

Let's look at some of images of chaos close-up. I'll use the program Fractals to explore various ways of looking at chaos. The non-linear equations that I will be dealing with in Fractals use complex numbers for their calculations. Complex numbers are numbers with both a real component and an imaginary component. The term *imaginary component* is actually a misnomer, as this part of the number is every bit as important as the real component. In complex numbers, you can think of the real portion as representing the amplitude of the

> **point attractor:** a specific point that a stable system tends to return to.

There are a number of examples of stable systems in the real world. The planets in orbit around the sun are examples of stable behavior where the attractor is elliptical (the orbit of a planet around the sun is an ellipse). A spinning gyroscope is stable with an attractor forcing the gyroscope to align itself to point toward the center of the earth. Many resource-based economic goods (such as farm products and minerals) tend to exhibit stable behavior in which the price and demand fluctuate in very predictable relationships with each other.

number while the imaginary part represents the *phase*. For wave phenomena, the phase represents whether the wave is at a peak, a valley, or somewhere in-between. All of the examples in Fractals divide the Fractal Window up into a real plane along the X axis and an imaginary plane along the Y axis. This allows you to select any point in the window and tell its real component (the X value) and its imaginary component (the Y value).

For all of the numbers represented in the window's complex plane, I look at the behavior of a simple non-linear equation with an order of magnitude of two when given that number as a starting point:

$$z = z^2 + c$$

Where z is a complex number and c is a complex number.

I iterate the equation while watching the behavior of Z. If Z begins to go toward infinity, I stop and color the dot on the screen. The color I use is based on the number of iterations that were needed before it went out of bounds. If Z stays stable, I leave the screen location black.

Let's think about what you might expect to see given this situation. You might expect to see a fairly clear demarcation point where the numbers began to grow toward infinity. This point will often be one, with numbers below one going toward zero and numbers above one going toward infinity. Another possibility would be banding, perhaps with even numbers going to infinity and odd numbers remaining stable. Because you are dealing with the complex plane, it might be reasonable to have one or more quadrants in the plane go toward infinity with the remainder going toward zero.

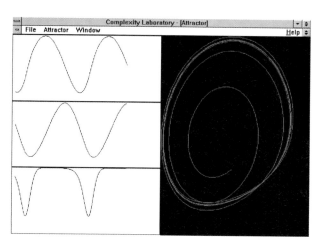

Figure 14.1.
Point attractor.

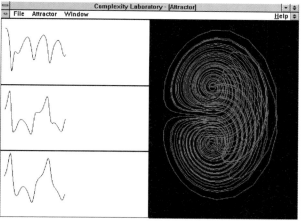

Figure 14.2.
Circular attractor.

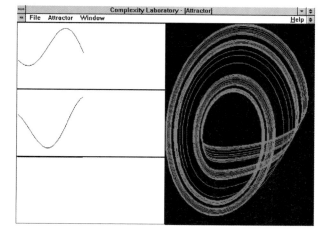

Figure 14.3.
Lorenz attractor.

complex number: a number with a real and an imaginary component. The square root of −1 is an imaginary number.

Figure 14.4.
Julian attractor.

Figure 14.5.
The initial Mandelbrot set.

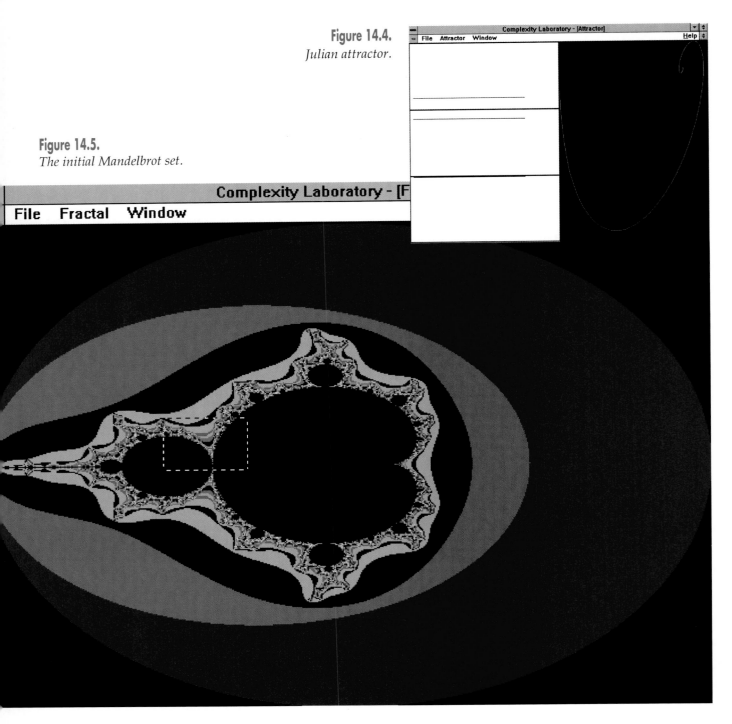

Let's observe what really does happen. Using a range of –2 to +2 for both the real and imaginary axis, a figure similar to Figure 14.5 is produced. The complex-looking structure that emerged is not at all what you expected. In fact, this structure is so surprising and unique that it is named after the first scientist to discover it. It is called the Mandelbrot set.

OK, so it's more complicated than expected when I use the default scale. As I zoom in, it should begin to simplify. After all, you wouldn't expect to see much difference in behavior for points scaled over, say, a .001 range. Figures 14.6 and 14.7 show the results of repeatedly zooming in on the Mandelbrot set (the dotted boxes in Figures 14.5 and 14.6 show the area being zoomed). In fact, you can zoom in on the Mandelbrot set to a degree limited only by the mathematical precision of your computer, and you will continue to see the same intricate structures. This is a visual portrait of chaos. Because it is a self-similar mathematical structure, it also is an extremely complex example of a fractal. It doesn't matter over how narrow a region you chose, you can continue to find points that go to infinity (colored) and those that remain bounded (black).

By changing the test for divergence from the Mandelbrot test to the Chaotic Curls test (a relatively small change in the equations), you can change the visual appearance of chaos significantly. Figures 14.8, 14.9, and 14.10 show progressive zooms into the world of chaotic curls. It is significant to note the ever-repeating complexity of these spiral structures. You can zoom in on spirals and never exhaust their detail and complexity.

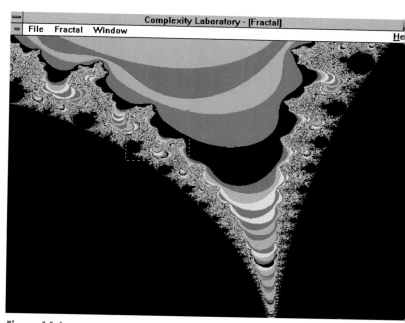

Figure 14.6.
Zoom-in on the Mandelbrot set.

THE TRANSITION FROM STABILITY TO CHAOS

> **cellular automaton:** a cellular program that generates future cells based solely on the conditions around it.

The program Tapestry is an example of a one-dimensional cellular automaton. Each row of colored dots is determined based on the color patterns of the dots in the previous row. For example, a rule might say, "If the previous dot color is blue, then the new dot color is red." Because the rules are expressed in terms of the previous color of this and neighboring dots, the program is called a cellular automaton. Because each new row is only dependent on the previous row (a line of dots), the cellular automaton is called one-dimensional.

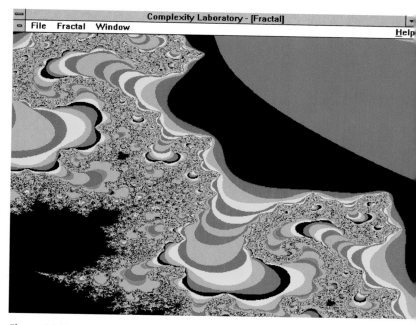

Figure 14.7.
A further zoom-in on the Mandelbrot set.

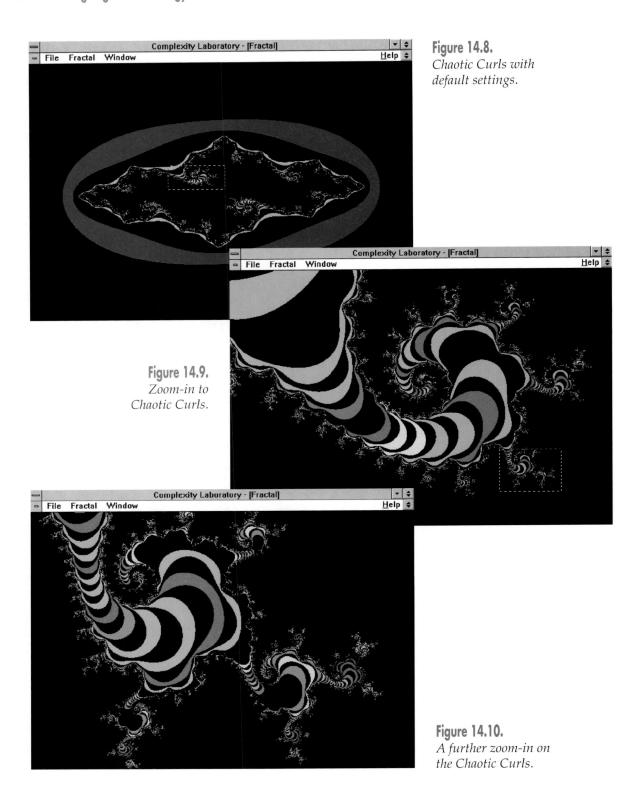

Figure 14.8.
Chaotic Curls with default settings.

Figure 14.9.
Zoom-in to Chaotic Curls.

Figure 14.10.
A further zoom-in on the Chaotic Curls.

It is interesting to look at the various states of the Tapestry cellular automaton over time. Tapestry does this by displaying the initial state on the top-most row of the screen, the second state on the second row, the third state on the third row, and so on. In effect, the screen shows the history of the cellular automaton with the higher rows being older versions while the lower rows are the newer versions.

Chaos and Complexity

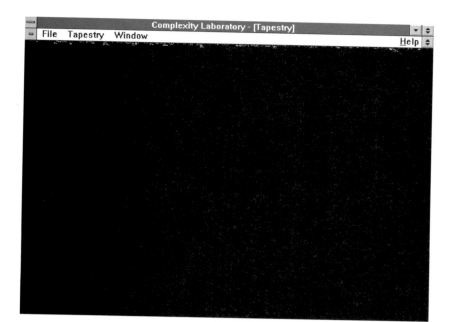

Figure 14.11.
Cellular automaton demonstrating stability.

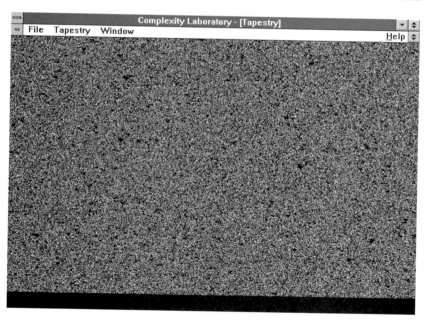

Figure 14.12.
Cellular automaton demonstrating chaos.

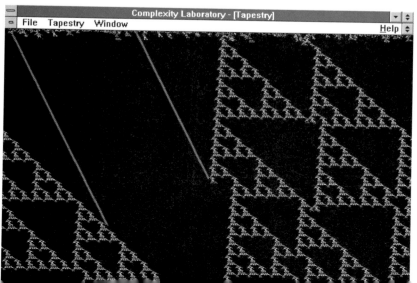

Figure 14.13.
Stable Tapestry pattern.

Figure 14.14.
A more complex, stable Tapestry pattern.

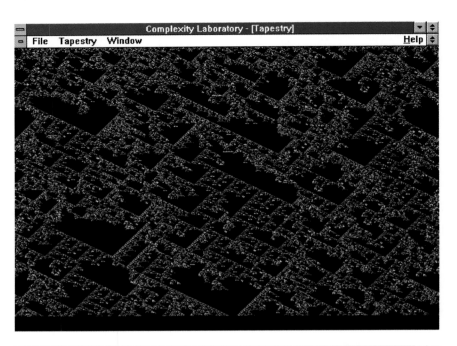

Figure 14.15.
Tapestry on the edge of chaos.

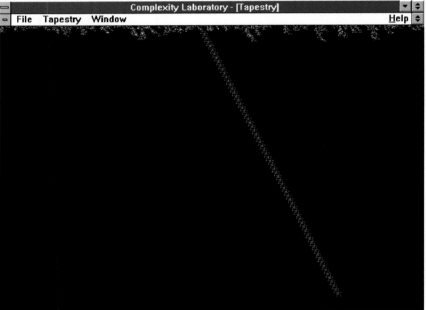

When working with cellular automatons, black cells are often called dead and colored cells alive. To simplify the process of playing with stability, complexity, and chaos, I designed Tapestry to automatically generate cellular automaton transition rules with a user-specified probability of the cell remaining alive. I called this factor the Z factor. Lower values of Z mean that the cells are more likely to die, whereas higher values of Z mean they are more likely to stay alive either by changing to a new color or by remaining the same color.

Figure 14.11 shows what happens with a low value of Z (.2). The cellular automaton starts in a random state (as shown by the colors at the top of the window) but quickly dies to a stable, dead state. Figure 14.12 shows what happens with a high value of Z (.75). The cellular automaton starts in a random state and remains chaotic. There is no order, or structure, in this system.

In the transition region between stability and chaos, interesting complex behavior can be observed. Figures 14.13 through 14.15 were produced with K varying from .42 through .5. Figure 14.13 is an example of an interesting, stable structure. The cellular pattern slowly moves itself from left to right. This pattern will repeat forever. Figure 14.14 is a good example of multiple repeating and stable patterns beginning to emerge. The cellular automaton in Figure 14.15, with a K value of .5, is clearly approaching chaos, but there are definite examples of underlying structures.

LOOKING FORWARD

Although the science of chaos and complexity are relatively new, they hold the promise of describing a wide range of physical phenomena that we observe whenever there are interacting systems. You can expect to see references to this important area of mathematics in such diverse areas as biology (complexity may explain cell differentiation, genetic triggers, and how memory works); evolution (complexity may explain how life emerged from the primordial soup); and economics (complexity may explain the behavior of economies, markets such as the stock market, and the dynamics of technology dominance). In addition, the theories of complexity have already resulted in practical applications in the areas of robotics and genetic programming.

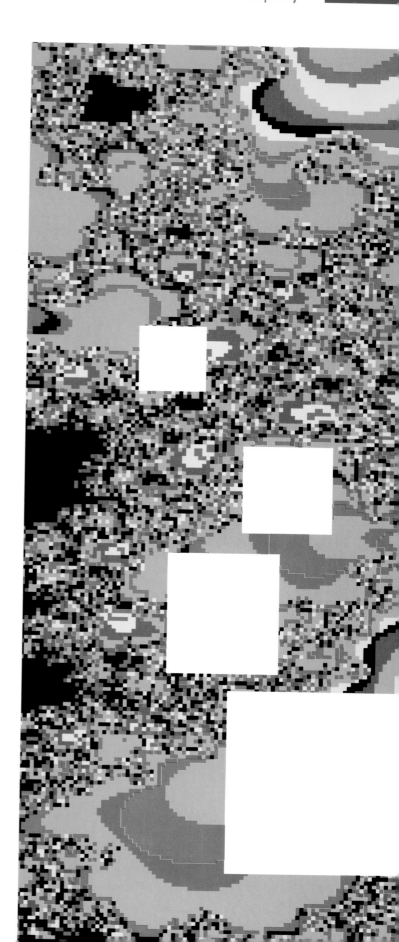

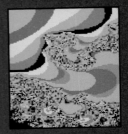

CHAPTER 14

Chaos and Complexity

bill roetzheim

Bill Roetzheim is currently working on *Enter the Complexity Laboratory*, a book devoted to chaos and complexity.

Glossary

brad jones

2-D 1 7 10 Having height and width but no depth; flat.

3-D 1 6 7 10 Having height, width, and depth.

3-D animation 1 6 12 A moving graphic picture that appears to have height, width, and depth.

3-D computer graphics 6 Graphics created in a computer that are not only three dimensional (similar to sculpture), but also can have lights with color and position, shadows, or even atmospheric effects such as rain and snow. In addition, motion can be added.

3-D graphics 7 The field of computer science concerned with representing objects in three dimensions on a two-dimensional screen.

3-D life 9 The study of life using a third dimension. See *life*.

3-D mouse 7 A mouse that can be moved forward and backward, left and right, and up and down. Some 3-D mice also can be tilted. Logitech of Fremont, California has a mouse that operates by controlling objects on x, y, and z axes as well as in pitch, roll, and yaw orientations.

3-D sound 13 The ability to hear sound coming from a direction whether real or perceived. The sound can appear to come from above, below, and beyond the edges of the speakers.

3-D space 6 A vast dark void within the computer that can be populated with a 3-D model.

3DO 12 A home game system that will try to bring the power of a high-speed graphics computer down to the price of a video game unit. 3DO will offer video games, full-length motion pictures that can be viewed directly from CD-ROM, and more.

4-D 1 Having height, width, and depth and also able to move interactively.

acoustically absorbent 13 Creating "dead" spots in sound; absorbing sound waves.

acoustics 13 The branch of physics concerned with the properties, production, transmission, and effects of sound.

actuators 5 Active components that can perform tasks. Typically part of a smart material.

agent 2 A character enacted by the computer that acts on behalf of the user in a virtual (computer-based) environment.

ambient light 6 The amount of background light in a room. The light is cast in all directions for a limited distance.

artificial intelligence (AI) 2 9 12 A computer science field that attempts to improve computers by endowing them with some of the characteristics associated with human intelligence, such as the capability to understand natural language and to reason under conditions of uncertainty.

artificial life programs 9 A cellular automation that is used to model biological organisms and ecosystems.

artificial life systems 2 See *life*.

artificial neural networks 2 Connected systems of simple processing units whose behavior is determined by the existence and strength of the interconnections between the units.

atomic manipulation 4 The ability to move atoms where you want them.

audio 13 The electronic representation of sound.

auditory system 13 The ear along with the brain.

automatons 14 Robots that move, speak, and generally behave in amazingly lifelike ways, yet each of their movements and gestures is preordained at the time their programs are written.

autonomous agents 2 Agents that do not employ human input in their decision processes.

backwards beam tracing 8 See backwards ray tracing.

backwards ray tracing 8 Ray tracing that starts from the source of the light rather than from the camera's viewpoint.

binary 11 A number system that uses a base of 2; 0 or 1; true/false.

binaural 13 The use of two sources to create or receive sound (corresponding to the use of two ears to listen).

blends 1 The combining of two or more images into one smooth image.

Boids 9 A computer animated program of bird creatures.

Bomb 9 A computer virus that is typically triggered by a date, time, or event.

boot-sector virus 9 A computer virus that destroys the directories and file allocation tables (FAT) of the hard drive.

bump maps 6 8 Texture maps that also apply an image to simulate texture in a 3-D object.

capillary waves 8 The smallest type of waves, caused by wind; these waves can be as small as a fraction of an inch.

caustic optical effects 8 The shimmering patterns on the bottom of a pool of water.

caustic polygon 8 A beam of light's footprint left when it strikes an object.

caustics 8 A term stemming from the burning of hot spots by a magnifying glass in sunlight. This refers to black spots created when rendering the interaction of light and water. Black spots are caused where all the light rays are reflected away from a spot.

CD-ROM 12 Compact Disc Read-Only Memory. Disks that contain information (such as video clips) that can be manipulated but not overwritten.

cellular automation (CA) 9 14 A program that uses rules that are expressed in terms of neighboring information. For instance the color of the current cell would be based on the previous color of the cell along with the neighboring cells' colors.

cellular automation transition rules 14 A cellular automaton that is dependent only on the previous row.

centroid method 11 A method of defuzzification that provides the output value with equal areas above and below a point.

chaos 10 11 14 A limiting behavior in which there is no correlation between the system behavior at various times.

chaotic 14 Operating in a seemingly random fashion with no patterns, no memory, and no emergent behavior.

chess machine 2 A small device that incorporates artificial intelligence, graph search, pattern recognition, and other techniques to play chess.

chromatic dispersion effect 8 The way a prism creates rainbows or a diamond flashes different colors.

class 11 A set of properties that describes a group.

classification (for reasoning) 11 The concept that a partial match to one or more classes can be made from a current case.

co-evolution theories 9 A newer theory of evolution that states organisms cooperate to survive as a species.

cocktail party effect 13 The ability to subconsciously filter out the louder sounds around us so that we can concentrate on the more quiet sounds.

combined-code offspring 9 The combining of two programs to produce a new program.

complex 14 As defined by Websters, "...a group of related ideas and activities, things, etc. that form, or are viewed as forming, a single whole."

complex model 14 A computer program that frequently behaves in surprising or unexpected ways.

complex number 14 Numbers with both a real component and an imaginary component. The real portion represents the amplitude of the number, while the imaginary part represents the phase.

complex transition region 14 The region between stability and chaos.

complexity 14 As defined by Websters, "...anything complex or intricate..."

complexity science 14 The study of emergent behavior exhibited by interacting systems operation at the threshold of stability and chaos.

computer worms 9 See *viruses*.

convolution (sound) 13 The filtering of an input signal by head-related transfer functions.

Core Wars 9 A recreational form of computer viruses that was created by A.K. Dewdney. The game allows two programs to battle and annihilate each other.

cross-screen filter 8 Filters used by a photographer to create starts on highlights.

current case 11 A particular set of observations.

cyberspace 2 7 The "place" created by the networking of multiple VR systems and environments; the information space in which we all operate (i.e. the place you are when you're having a phone conversation).

data glove 7 A device that collects the user's hand movements for use in a virtual reality system.

defuzzification 11 A collapsing of an output fuzzy distribution into a single value.

desktop Virtual Reality 7 A low-cost method of interacting with a virtual world using a standard personal computer. The monitor serves as a "window" into the virtual world.

deterministic 10 Determined by a sequence; (fractals) perfectly pure mathematical entities such as the Mandelbrot set.

diffracting 8 The breaking of light rays into dark and light areas or into color spectrums.

diffraction effect 8 Shadows in an image caused by light being diffracted.

digital 1 4 Computerized.

digital puppet 1 A computer-generated character that can be controlled by external devices such as a joystick, hand glove, mouse, or keyboard.

digitize 6 A method of modeling in which a laser is shined onto the surface of the object while sensors are used to detect the variations as the laser pans across the surface. Commonly used to create models of actors such as those in the movie *Terminator 2: Judgment Day*.

dimension 10 Measurements of length (or height), width, depth, and/or time.

dissolve 3 Also called cross-dissolve or lap-dissolve, this is Hollywood-speak for fading out one image while fading in another. Because the fades overlap, the screen intensity doesn't change and the effect is very smooth. It is distinct from the straight fade, where the screen goes to black. The computerized version uses the following method: for the first tween of a ten frame sequence, take ten percent of the source image and ninety percent of the target image and combine them. For the second tween, add twenty percent of the source with eighty percent of the target, and so on.

distant light 6 Lights similar to spot lights except that the rays are all parallel instead of conical.

Dolby 13 A specialized sound system that uses more than two speakers.

Domains of attraction 10 Quaternion shapes.

doppelganger 1 A figure on the computer screen that matches the controlling person's movements.

dumb material 5 A material that cannot exhibit any intelligence or responsiveness.

e-mail 2 Electronic (computerized) mail sent from a user on one computer network to another user on the same or a different computer network.

edutainment software 12 Computer games or software that not only entertain, but are also educational.

electronic DNA 9 The binary values of a program's code.

embedded shape memory alloys 5 See *shape memory metal*.

emergent behavior 14 Unexpected, non-programmed behavior that emerges as a program runs.

emitters 1 Devices that emit a magnetic field that receivers can sense. Often used for tracking.

endomorphic anticipatory agents 2 Agents that use models of themselves and others in order to make predictions about future actions.

environment maps 6 Texture maps that are similar to reflection maps except that the reflected image is created at the time the computer renders the object.

extruding 6 See *lofting*.

fault diagnosis 11 Classifying observations into classes that are the faults to be detected.

FAX back system 2 12 A system that provides the capability to send FAXes (such as product information or other images) to the user automatically in response to the user's interaction with the system.

flight simulator 12 A program or machine that provides the illusion of flying.

fluid simulation 8 Computer representation of the way fluid moves and interacts.

flying mouse 1 See *3D mouse*.

focal point 8 The bright dot where all the light rays come together.

focusing 11 Adding considerations to potential options with some measure priorities in order to maximize discrimination between options.

fourth dimension 10 (fractals) A 90 degree angle "perpendicular" to our physical plane.

fractal 4 6 10 14 A curve or shape that repeats itself at any scale.

fractal algorithms 6 A very popular way of creating natural, organic-looking geometry.

fractal formula 6 Formulas used by the computer to generate random, natural-looking objects, such as mountain ranges, rocks, planets, plants, and trees.

fractal geometry 10 Geometry that is self-similar.

fractional dimension 10 The area between the second dimension and the third dimension.

frequency attenuation 13 The process by which the ear's shape slightly changes certain frequencies of the sound wave, depending on the direction the wave comes from.

full-body animations 1 Animations that are created using the full body movements of the subjects modeled from. Many animations only capture the movement of certain parts of the body such as the face or hand.

fuzzification 11 Relating observations into fuzzy values.

fuzziness 11 See *imprecision*.

fuzzy classifier 11 Information that is used to establish the terms that will be used to determine which fuzzy value to use.

fuzzy logic 11 A mathematical based scheme for modeling and controlling complex systems by recognizing that complexity and precision are generally contradictory; the more complex a system is, the less precisely we are able to compute its behavior from measurements of various inputs and conditions in the environment.

fuzzy set 11 A set that contains members that might only partially meet the set's criteria.

fuzzy variable 11 A variable that can have a range of values.

game-playing agents 2 Agents that play games such as the chess machines. Some have specialized applications such as economic analyses and military simulation.

garage VR 7 See *desktop Virtual Reality*.

genetic algorithms (GA) 9 A class of artificial life programs that evolve in a Darwinian, survival of the fittest, manner. These programs are usually problem-specific.

geometric primitives 6 The basic geometric shapes that are created with simple mathematical formulas. (For example, spheres, cubes, and cylinders.)

graininess 11 A level of measurement that mimics actual knowledge or measurements that are not always smooth.

graphical user interface (GUI) 7 A computer interface that uses pictures and intuitive symbols instead of textual information.

hard-edged classifier 11 Classifiers that can have only certain values (such as off, slow, or fast).

head mounted display (HMD) 1 7 A device worn on the head that may or may not contain monitors. It allows the wearer to view 3-D images. In addition, most HMDs have trackers that can register the movement of the head.

head tracking 13 Using a tracker to monitor head movements.

head-related transfer function 13 The reflection of sound off our shoulders, the distance between our ears, and the shadowing effects of the head.

Heuristics 11 Rules of thumb.

HMD 1 See head mounted display.

home game system 12 A machine designed to allow the user to play interactive games.

homebrew VR 7 See *desktop Virtual Reality*.

human computer interface 7 A device that allows a person to interact with her environment (either real or virtual).

hyperlink 12 To connect to another product, episode, or program that has a connection with the product, episode, or program currently being used in order to access additional information.

illumination map 8 Similar to a texture map, this is a map that contains the light patterns based on light reflections/refractions.

ILM 3 Industrial Light and Magic, a division of LucasArts in San Rafael, California. This is the computer graphics studio that started it all. Look at the end of any recent movie with great special effects and you will see a credit for ILM. These guys are the champs.

immersion 2 7 To experience a computer-simulated world as if you were a part of it or within it; the degree of the quality of the visual, aural, and tactile rendering as well as the lags inherent in the tracking and display system.

immersive VR 7 A feeling of being a part of (or inside) the virtual world.

infected (diskette) 9 A diskette or program with a computer virus or worm.

Information America on-line 2 A computer database that cross indexes the Postal Services' National Change of Address file, subscription lists, birth records, driver's license records, telephone books, voter registration records, and records from numerous government agencies.

intelligent agents 2 Specialized software tools that are employed to perform tasks for users in virtual reality; entities which reside in virtual environments.

intelligent structures 5 See *smart structures*.

interactive (interaction) 2 7 (virtual reality) The virtual world responds to your actions.

interactive entertainment 12 Products that provide entertainment that allow the user to participate (for example, coin-operated video games and computer entertainment software).

interactive services 12 Services to be provided by either the cable company or phone company that allow the user to access options interactively.

interaural crosstalk 13 The crossing of sound. When sound from the left speaker reaches the right ear and sound from the right speaker reaches the left ear, causing smeared sonic imagery.

interpolates (interpolation) 1 2 The calculation of all the positions between the key frames.

Inverse Kinematics 6 An animation method for simulating falling objects. The user simply specifies that the object should fall, the author specifies how much gravity the computer should simulate, and the computer handles the rest of the animation.

Julia sets 10 Convoluted fractal forms whose twists and turns remind one of the scales of a sinuous dragon.

key frame 1 6 A single frame or image that may appear in an animation.

Lathe modeling 6 A method of modeling similar to lofting; however, instead of stretching a two dimensional object, it is revolved around itself to create a three dimensional object.

left-hand-size (LHS) condition 11 Input; the value on the left side of the equal sign in an equation; the observed object or case.

life 9 A relatively new field of study which is attempting to understand the basic principles of living organisms and ecosystems through the use of computational models and systems which exhibit these behaviors.

localize (sound) 13 The ability of the brain to use time/phase difference, loudness difference, and frequency attenuation to determine the exact location of a sound's source.

lofting 6 A method of modeling where a two dimensional image is stretched into a three dimensional image.

loudness difference 13 What occurs when studio engineers pan sounds to create a conventional stereo image.

M-set 10 See *Mandelbrot set*.

Mandelbrot set 10 14 The most common fractal, named after "the good doctor"; the sum of all the Julia sets (of which there are an infinite number—each of which is infinitely complex).

mating 9 The process of creating a combined-code offspring.

max-dot 11 A procedure that multiplies each point of the output membership function by the LHS truth value.

max-min 11 A procedure that clips the output membership function at a level corresponding to the LHS truth value.

mean of maxima 11 A defuzzification method that provides a weighted average of the points with the same highest membership.

mechanical chemistry 4 Similar to solution chemistry; however, instead of working with bulk chemicals, molecules are used one at a time using nanoscale robots to chemically "grab" them.

mechanosynthesis 4 See *mechanical chemistry*.

membership 11 Belonging to a set.

membership function 11 A single-valued function used for fuzzification. For controller applications, a simple triangle or trapezoid has proved effective.

membership value 11 Values derived for an observation by using fuzzy logic.

metaballs 6 A technique of simulating natural phenomena by using drops of liquid that an artist can enlarge, stretch, and fuse together with other drops (metaballs) to form a naturally curving surface.

micro- 4 One millionth.

micromachine 4 Tiny motors and machines that sound nanotechish, but are a billion times larger in volume than a nanomachine. These machines are built the same way that other things are built—by carving, sculpting, or casting—not by mechanosynthesis.

milli- 4 One thousandth.

MIPS 9 A computer measurement—Millions of Instructions Per Second.

modeling 6 8 9 The process of creating 3-D objects within the computer by instructing the computer to make simple geometric shapes such as a sphere, cube, cylinder, and so on. Complex objects are typically created by combining many of these geometric primitives.

molecular machinery 4 Machines built at the molecular level that can be used for any purpose, including building additional molecular machines.

molecular manufacturing 4 Making objects molecule-by-molecule with molecular machines.

mono 13 Coming from a single sound source.

morphing 3 Shorthand for metamorphosing; it refers to a sequence of smooth changes between two different images, such as a man and a werewolf, or a bat and a vampire. Technically, morphing combines a warp and a dissolve.

motion-capture system 1 A system that uses motion sensors to capture movement for animation.

motion sensors 1 Sensors used to capture the movement of a device and translate it to the computer in order to be used with animations.

multimedia 2 12 The presentation of information on a computer using graphics, sound, animation, and text.

Musical Instrument Digital Interface (MIDI) 12 A standard interface for musical devices.

mutate 9 The random changing of a program's code in order to mimic evolution.

nano- 4 One-billionth; from the Greek for dwarf.

nanolithography 4 Arranging atoms to form letters and words.

nanoscale robot 4 A super-small device that will be able to grab molecules to be used in mechanical chemistry.

nanotechnology 4 The creation of microscopic machines.

natural language interface 2 An interface that allows simple English sentences and scripts to be used.

natural model 5 A model that is not man-made (for example, skin).

nervous system 5 The part of a smart material that senses what is happening within or nearby the structure.

non-deterministic 10 Determined with some random elements; (fractals) mathematical entities that have a random element added in order to simulate images of things with the organic beauty of natural forms.

non-linear storyline 12 Storylines that can change based on selections made.

nonlinear systems 11 Systems with long-term behavior that is independent of how the system starts.

omnidirectional (sound) 13 See *3-D sound*.

optical system 1 A system that uses light or light sources.

orbitals 8 The circular paths that points in water travel along.

organism 9 A name for program code in a life system.

particle system 6 A collection of very tiny particles that move in any direction specified. They are used to simulate natural phenomena such as splashing water, snow, sun flares, volcano eruptions, and more.

PDI 3 Pacific Data Images, Los Angeles, California. This company is the primary competitor to ILM (Industrial Light and Magic) in the computer special effects department. The warping program on the included disk is based loosely on the algorithm developed at PDI to do the Michael Jackson video, among others. Very clever people.

pencil 8 An optical term referring to a small bundle of rays.

performance animation software 1 Software that allows an on-screen digital puppet to be controlled using any number of real-time input devices such as a mouse, joystick, motion-capture system, or the keyboard.

photochromic glass 5 Glass that changes colors based on light entering it and causing a photochemical reaction that coerces silver ions mixed into the glass to temporarily extract an electron from nearby copper ions, and therefore cause the glass to darken.

piezoelectric 5 A sensor material that relies on electricity; a word derived from the combination of the Greek word meaning to press and the word electric.

pinna 13 The shell-like shape of the outer ear.

pixel 12 A single dot of light.

plan 2 A sequence of actions that produces a solution to the problem of interest.

point attractor 14 A point that a system tends to return to such as the straight-down position of a clock's pendulum.

portable game unit 12 A small game unit that can be carried easily.

projection 6 See *rendering*.

projection Virtual Reality 7 A form of virtual reality that resembles Desktop Virtual Reality, but the window into the virtual world is much larger and provides a better view for many people or a wider view for one person. The user sees an electronically generated image that is controlled by moving around inside of a controlled space.

proximal probe 4 Probes that are small enough to move atoms.

psychoacoustic 13 The branch of psychophysics that explores how we perceive and interpret sound.

quadraphonic 13 Four-channel sound.

qualitative 11 Measurable in quality or quantity.

quaternions 10 The technique of 4-D algebraic operations.

radiosity 6 8 A technique that simulates the way light bounces and reflects off of other objects.

randomness 9 Factors that allow accurate modeling of weather patterns, population growth, and the spread of infectious disease in populations.

ray tracing 6 8 A method of coloring a 3-D object's surface that traces every ray of light in a 3-D scene from the viewpoint of the camera through the scene to the ray of light's origin. This is the opposite of what happens with the human eye, where every ray of light starts from its origin and bounces off objects and into the human eye.

real-time 1 2 Occurring instantaneously.

receivers 1 Sensors used to track the position of a person as they move through a magnetic field.

reflection 8 The bouncing of light off of a surface.

reflection maps 6 A texture map that takes a single image and reflects it off the surface of the object. If the camera moves around the object, the reflection shifts realistically.

reflective surfaces 6 See *reflection maps, environment maps,* and *ray tracing*.

refraction 6 The effect of light shining through a solid object or a liquid.

render 1 10 To recompute at higher resolution with more complex shading.

rendering 2 6 8 Using light and shadows to make a computer-generated model look realistic.

retroreflective 1 Reflective inward rather than off of a surface.

right-hand-side (RHS) action 11 Output; the actions that will be applied to the input, or left-hand conditions.

rules (artificial life) 9 The set of conditions that determine how an artificial organism lives, dies, and reproduces.

rules (fuzzy logic) 11 The terms that will be used to classify observations into classes. Typically of the form IF <conditions> THEN <class>.; a mathematical relation between the left-hand-side (LHS) conditions and the right-hand-side (RHS) actions.

scanning tunneling microscope (STM) 4 A device that has a tip so sharp and positioned so exactly that it can be used to move atoms.

self-organizing artificial life 9 Life programs that exhibit evolving forms and structures.

self-organizing system 9 Systems that form their own organizational structure such as bromate ions in a high acidic medium that create concentric, circular rings and spirals.

self-similarity 10 Statistically the same; unvarying under the change of scale.

set 11 A collection of things specified by defining how to determine whether a thing belongs to the set (i.e. "all integers between 6 and 10" produces the set 7,8,9).

shaded model 1 An image that has a surface applied so that it appears solid.

shape memory metal 5 Metals that can be bent into some form, but when heated beyond a transition temperature, they will try to assume a shape that they had been in when last at that temperature; A metallic act of remembering.

SIGGRAPH 1 A popular computer graphics conference sponsored by the Association for Computing Machinery's Special Interest Group on Graphics.

signal processing 13 A technique used in the studio to change the characteristics (such as pitch) of an audio signal.

Simulation Virtual Reality 7 The oldest form of virtual reality. This method uses a cabin and video screens or computer monitors to provide windows to a virtual world. The cabin includes realistic physical controls.

simulator sickness 7 Nausea that can accompany extended virtual reality experiences.

Stealth virus 9 A virus that can be either a boot or file virus. This virus tries to hide from anti-virus programs.

smart materials 5 Materials that appear to be intelligent in that they are sensitive to, or can respond to, changes in their environment (such as photochromic glass).

smart structures 5 Items made from smart materials.

solution chemistry 4 A bulk technology that works with statistical populations of molecules by sloshing liquids around in test tubes.

sonic 13 Dealing with sound.

sonic imagery 13 The mental images created by sound.

sound waves 13 Regular vibrations of air that create sound.

source 6 Where a light or camera originates from.

spatialized sounds 13 See *3-D sound*.

spatially enhanced stereo 13 Another name for 3-D sound.

speech recognition technology 2 Technology that enables a user to have limited spoken interactions with intelligent agents controlling voice mail, FAX back, and other systems.

spot light 6 A light that has both a source and a target, with a cone of light of a specified width. This is similar to an invisible flashlight that can be positioned and pointed in any direction.

stable system 14 A system that both exhibits repetitive behavior and that tends to return to this behavior when artificially forced to a different behavior.

stereo 13 Using two sources to either record or play back sound.

stereophonic 13 Sound that was recorded with two or more microphones and played back with two or more speakers.

stereoscopy 7 The viewing of objects so as to appear three-dimensional.

"Stoned" virus 9 A boot-sector virus that displays the message "Your PC is stoned—LEGALIZE MARIJUANA."

strange attractors 14 Attractors that do not correspond to a simple geometric shape.

Surface of Revolution modeling 6 See *Lathe modeling*.

texture maps 6 2-D images that the computer wraps around a 3-D object.

The Elite Motion Analyzer 1 A 3-D computer animation application produced by Bioengineering Technology and Systems (Milan, Italy). This system uses video cameras with infrared light sources around the lens, captures the x and y positions of retroreflective markers and then calculates the z position.

the Oz project 2 A virtual world created at Carnegie Mellon University in which four "blobs" interact with each other in a an animated environment.

the Z factor 14 A user defined probability of a cell remaining alive.

THX sound 13 A special sound system that uses more than two speakers.

tidal waves 8 The largest type of wave, caused by geological forces such as earthquakes.

time/phase difference 13 The effect caused by the fact that sound waves don't strike both ears at precisely the same time.

tracker 1 7 A device that can sense movement and translate it to the computer.

trends 11 Directions and magnitude of the change in values in a system with a fuzzy controller.

truth value 11 The value obtained by evaluating each of the membership values from member functions.

tween 3 Shorthand for in-be*tween*ing. It refers to a series of images that smoothly connect two other images. If the images are similar—say a horse in two different poses—the result is regular animation. If the images are different, tweening creates a smooth metamorphosis called morphing.

universe of discourse 11 The set of ranges that the observed cases of a fuzzy-logic system can produce.

V-art 7 Art created in a virtual world environment.

VActors 1 See *virtual actors*.

ventriloquism 13 The practice of speaking so that the voice seems to come from some source other than the speaker.

video phones 2 Advanced phone systems that allow the use of recorded video images as well as voice interaction.

virtual 1 2 7 Existing only in the computer.

virtual actor 1 2 A computer-generator character or object controlled in real-time by people; a digital model of famous person, interesting person, or just about any other person that can be created using the technology of artificial agents.

virtual camera 6 A non-existent "camera"; based on the location of the camera, the computer will generate an image that represents the 3-D scene as it would be seen from the location of the camera.

virtual puppet 2 Agents completely controlled by a human operator.

virtual reality 1 4 6 7 12 13 A technology that lets a person enter, manipulate, and travel through computer-generated, interactive, three-dimensional environments: "virtual worlds".

virtual worlds 2 7 Environments created in a computer.

viruses 9 Self-replicating programs that inject themselves into computer operation systems. They have the ability to infect and replicate themselves inside the host files or diskettes.

viscosity 8 The internal thickness of a liquid that determines how quickly or slowly it will react.

voomies 7 Virtual movies in which the audience interacts with the movie actors.

waldo 1 The generic name for a control device that mimics the things it's controlling.

warp 3 A technique for stretching and squashing an image.

water transport 8 The movement of water.

wave arrival time 13 See *time/phase difference*.

wireframe 16 A framework of lines that outlines an object (typically three dimensional). It is less work for the computer to manipulate an outline of an object than an actual surface model of an object.

witness points 1 Markers that can be tracked in order to record movement to be added to animations.

Index

Index

SYMBOLS

2-D, 159
3-D, 34, 70, 92, 131, 143,
　graphics, 60-68, 74, 159
　mouse, 77, 159
　sound, 138-139, 141-146, 160
　space, 107, 160
3-D Life, 95
3D Studio, 23
3DO game, 128, 160
4-D, 107, 110-112, 160, 163
16-bit Sega games, 128

A

The Abyss, 33-35
acoustics, 160
actions
　left-hand-side (LHS), 121
　right-hand-side (RHS), 121, 167
actuators, 51, 54-55, 160
agents, 160-163
　intelligent, *see* intelligent agents
AI, *see* artificial intelligence
aircraft,
　fatigue-sensing, 53-55
algorithms
　fractal, 63, 163
　genetic, 163
ambient light, 65, 160
analog systems, 42
animation, 2-7, 13-14, 23, 60-61, 131, 159, 163
　morphing, 28-30
　technologies, 60-67
　water, 82-92
antiviral
　software, 103
Appel, Art, 89

Armistead, William H., 50
artificial intelligence (AI), 24-25, 97, 160
　programs, 16-17
artificial life, 94-99, 100-103, 160, 168
artificial neural networks, 160
artificial reality, 74
artificial realities, creating, 60-68
Association for Computing Machinery Special Interest Group on Computer Graphics, 4, 29
atomic manipulation, 45-46, 160
attenuation, 140
　frequency, 163
attractors, 150, 169
audio, 160
auditory system, 160
automata, 148, 153-157, 160-161
autonomous agents, 17, 160
autonomy continuum, 17

B

backwards ray tracing, 89-90, 160
Bates, Joseph, 21
BattleTech Center, 77-79
beam tracing, *see* ray tracing
behaviors
　emergent, 149, 162
　repetitive, 150
　system, 118
Belousov-Zhabotinski reaction, 95-96
binary logic, 119
binaural, 160
　recording, 142
Binnig, Gerd, 45
blends, 160
blobby water droplets, simulating, 87-88

body, complexity theory, 148
boids (artificial life), 95, 160
Bomb virus, 160
boot-sector viruses, 160
Brigham, Tom, 29, 32
bump maps, 65, 160

C

capillary waves, 82, 160
Carpenter, Loren, 109
CAs (Cellular Automatons), 94
cases, 119
caustics, 88-90, 161
CD-ROM, 128, 161
Cellpro program, 100
cells, 148
Cellular Automatons (CAs), 94, 153-157, 161
Center for Intelligent Material Systems and Structures, 51-52
centroid method, 122, 161
chaos, 118, 150-157, 161
Chaotic Curls test, 153-157
Character Shop, 11
chemistry, 44, 165-168
chess machine, 161
Chiarini, Francesco, 5
chromatic dispersion effect, 161
class/action relationships, 120
classes, 119, 161
classical artificial intelligence, 24-25
classifiers
　fuzzy, 120, 163
　hard-edged, 164
co-evolution theories, 161
cochlea (ear), 140
cocktail party effect (hearing), 141, 161

Cohl, Emile, 28
collision avoidance
 controllers, 122
combined code
 offspring, 96-97, 161
complex models, 161
complex systems,
 118-123
complex transition
 region, 161
complexity
 science, 161
complexity theory,
 148-150
computer animation
 technologies, 60-67
computer graphics,
 3-D, 60-68, 73,
 109, 159
computer morphing,
 28-38, 166
computer worms, 161
computer-generated
 characters, 10, 13-14
 puppets, 5-7
concrete, self-
 repairing, 50, 56-57
controllers
 collision,
  avoidance, 122
 fuzzy, 120-123
controls, system, 119
convolution, 161
convolvotron, 145
Coons, David, 106
Core Wars, 103, 161
corrosion-sensing
 materials, 56-57
Crawley, Edward,
 51, 58
cross-dissolve, 162
cross-screen
 filter, 162
current case, 162
curved mirrors, 88
cyber-secretary
 agents, 19
cyberspace, 73, 162
Cyberspace
 Developer's Kit
 (Autodesk), 78
Cyberware, 34
cycloids, multiple,
 90-91

D

Dactyl Nightmare
 game, 75
damage sensors, 54
data bodysuits, 70
data gloves, 70-72, 75,
 80, 162
databases, 21-22
defuzzification,
 120-124, 162
deGraf, Brad, 3-10
depth cuing, 90-91
desktop Virtual
 Reality, 76-77, 162
deterministic
 fractals, 110
Dewdney, A.K.,
 103., 161
diffracting, 162
diffuse light, 89
Diffusion Limited
 Aggregations (DLAs),
 115
digital, 162
digital puppets,
 3-7, 162
digital systems,
 42-43
digitizing, 162
 3-D shapes, 62-63
dimensions, 112, 162
 fractal, 113
Dippe, Mark, 34
diskettes,
 infected, 164
dissolve, 162
distant light, 65, 162
DLAs (Diffusion
 Limited
 Aggregations), 115
DMorph (Dave
 Mason), 36
DNA, electronic, 162
Dolby, 162
Domains of Attraction,
 113, 162
doppelganger, 162
Drexler, Eric, 42-47
Dry, Carolyn, 56
Dryver, Huston, 53
dumb material, 162
dummy head record-
 ing, 141-142, 145

E

e-mail, 162
ear, 139-140
earthquake sensors,
 52-56
Edutainment
 software,
 130-134, 162
El-Fish program, 100
electronic DNA, 162
The Elite Motion
 Analyzer, 5, 169
ellipsoids,
 primitive, 87
embedded shape
 memory alloys, 162
emergent behavior,
 148-149, 162
emitters, 162
Emme, L., 109
endomorphic
 anticipatory
 agents, 162
energy, waterwave,
 82-84
entertainment, 126
 interactive,
  127-128, 164
 virtual reality, 79
 see also Edutainment
environment maps,
 66, 163
environments,
 3-D, 70
ethics, virtual
 reality, 80
evolution, 98
extruding, see lofting

F

FastTract device, 13
FAT (file allocation
 table), 102
fatigue-sensing
 materials, 50, 53-55
fault diagnosis, 119,
 122, 163
fault-detection
 systems, 123
FAX back system, 163
Feynman, Richard, 42
fiber optics, 53-56
file allocation table
 (FAT), 102

filters,
  cross-screen, 162
Fitch, Charles, 108
flight simulators, 163
fluid simulation, 163
flying mouse, 163
focal point, 163
Focal Point system, 145
focusing, 119, 163
food sharing, 98
formulas, fractal, 163
Fournier, Alain, 84-86
fourth dimension, 107, 110-112, 163
fractals, 63, 106-116, 150-153, 163
fractional dimension, 163
frequency, 140
  attenuation, 163
Fuhr, Tom, 53, 56
full-body animations, 163
functions
  head-related transfer, 164
  membership, 121-124, 165
Furness, Tom, 75
Fusco, Mike, 11-12
fuzzification, 121, 163
fuzziness, 163
fuzzy classifiers, 120, 163
fuzzy controllers, 120-123
fuzzy diagnostic applications, 124
fuzzy logic, 118-119, 122-124, 163, 168
fuzzy sets, 118, 122-123, 163
fuzzy values, 121-123
fuzzy variables, 120-123, 163

G

GA (genetic algorithms), 163
Game Boy (Nintendo), 129
  Monopoly, 131
Game Gear (Sega), 129
Game of Life, 94-95
game-playing agents, 18, 163
garage VR, 76, 80, 163
Gardiner, Peter, 50-52
genetic algorithms, 94-97, 163
geometric shapes, creating, 62-63
geometry, fractal, 115, 163
Gibson, William, 73
glass,
  photochromic, 50
Glenn, Steve, 11
goggles, data, 75
graininess, 164
graphical user interface (GUI), 70, 164
graphics, 60-68, 73-74, 109, 128, 159
grid-pencil-tracing, 89-90
Grolier Guides project, 21

H

hard-edged classifier, 164
Harold Friedman Consortium, 5
Hart, John, 110, 113
head tracking, 164
head-mounted display (HMD), 70-75, 164
head-related transfer function (hearing), 141, 164
headphones, 141-142
hearing, 139-141
  see also sound
height fields, 85
Heilig, Mort, 73
Henson, Jim, 3-5
Heuristics, 164
Hillis, Danny, 97
homebrew VR, 76, 164
Hornet-1 flight simulator, 134
hot spots, 89
Howard, Ron, 30
Hu, Lincoln, 34
human-computer interface, 70, 73, 164
Huston, Dryver, 56
hyperlink, 164

I

ICOM Simulations, 131-132
IF THEN statements, 121
illumination maps, 89, 164
ILM (Industrial Light and Magic), 34-35, 164
images, 3-D, 60-61
IMM (Institute for Molecular Manufacturing), 45-46
immersive VR, 75, 80, 164
Indiana Jones and the Last Crusade, 33
Industrial Light and Magic (ILM), 29, 32, 164
infected diskettes, 164
Infocom, 131
Information America on-line, 164
information retrieval agents, 21
infrastructures, utilizing smart materials, 55-56
Institute for Molecular Manufacturing (IMM), 45-46
instruments, orbiting, 56
intelligence, artificial, 160
intelligent agents, 16-26, 164
interactive entertainment, 127-135, 164
  interfaces, 70, 73, 164-166
interactive personalities, 10
interpolates, 164
Inverse Kinematics, 67, 164

J

JACK system, 22
Jacobson, Linda, 80
Jerusalem-B
  virus, 103
Julia sets,
  107-108, 165
Julian attractor, 150

K

Kajiya, James, 89
Kass, Michael, 84-85
Keele, Steven, 12
key frames, 165
  specifying, 67
keyboards,
  MIDI, 130
Kleiser, Jeff, 3
Knoll, John, 34
Kolb, C. E., 109
Kory, Michael,
  5-7, 13
Kosko, 124
Krueger, Myron,
  74-75
Krummenacker,
  Markus, 46

L

L-systems, 109
Lanier, Jaron, 75
lap-dissolve, 162
Lathe modeling,
  63, 165
Lazzari, Umberto, 5
Leach, Geoff, 46
left-hand-side (LHS)
  actions, 121, 165
lighting, 88-91,
  160-162, 168
  techniques, 65
Limited Intelligence
  Agent (LIA), 22
live video, 131
localizing sound,
  140-141, 165
lofting, 63, 165
logic, fuzzy, 124
Lorenze
  attractors, 150
loudness difference,
  140, 165

Lucas, George, 29-30
LucasArts, 29-32
Ludwig, Carl, 87,
  90-92
Lynx (Atari), 129

M

M-set, see Mandelbrot
  set
MAGI/
  Synthavision, 90-91
Magic Edge
  company, 134
Mandelbrot, Benoit B.,
  107-109, 113-115
Mandelbrot set,
  108-110, 114,
  152-153, 165
manipulating,
  atoms/molecules,
  45-46, 160
manufacturing,
  molecular, 43-47, 166
maps
  bump, 65-66, 160, 163
  illumination, 89, 164
  reflection, 66, 167
  texture, 65, 91, 169
materials
  dumb, 162
  smart, 50-58, 168
mating, 165
Max, Nelson, 84
max-dot, 121-122, 165
max-min, 121-122, 165
mean of maxima,
  122, 165
Measures, Raymond,
  53, 58
mechanical
  chemistry, 165
mechanosynthesis,
  44-46, 165
media, static, 60
membership, 165
  functions,
    121-124, 165
  sets, 118-119
  values, 165
Merkle, Ralph, 46
metaballs
  technology, 64, 165
metals, shape-
  memory, 55

metamorphosing, see
  morphing
Metamuse, 97-98
methods, centroid, 161
micromachine, 165
microprocessors,
  embedding, 51, 58
MIDI (Musical
  Instrument Digital
  Interface), 130, 166
Miller, Gavin, 84-85
Minsky, Marvin, 42
mirrors, 88-91
mixing sound, 139
modeling
  Lathe, 63, 165
  techniques, 62-65
  versus
    rendering, 82
  water, 84-88
models, 161, 168
molecular
  manufacturing,
  43-47, 166
molecules,
  manipulating, 45-46
mono, 166
Monopoly (Game
  Boy), 131
Morph (Gryphon
  Software), 36
morphing,
  28-38, 166
Morphing Magic
  (Pacific Data Images),
  37-38
Motion Analysis,
  3-4, 14
motion sensors, 12, 166
motion-capture
  systems, 9, 166
mouse, 77, 159, 163
movies, 126
  computer morphing,
  32-36
Mr. Film, 12
multimedia, 166
multiple cycloids,
  90-91
Musgrave, F. Kenton,
  109, 114
music, 8, 97-98
Musical Instrument
  Digital Interface
  (MIDI), 166

mutate, 166
mutation (artificial life), 97

N

Nakamae, Eihachiro, 86, 91
nanolithography, 166
nanoscale robots, 166
nanotechnology, 42-48, 166
NASA Virtual Interface Environment Workstation, 75
  warping, 29
National Computer Security Association, 103
natural language interface, 166
natural models, 166
natural phenomena, simulating, 63-64
NEC Turbo Express, 129
nervous system, 166
networks
  fiber optic, 55-56
  neural, 160
neurons, 148
Nintendo Super Mario Brothers, 11
Nintendo Game Boy, 129
  Monopoly, 131
non-deterministic, 166
  fractals, 110
non-linear systems, 118, 166
normals, perturbed, 91-92
Norton, Alan, 112-115
nouvelle AI (artificial intelligence), 24-25
Nowak, Martin, 98

O

objects, 3-D, 62-65
omni light, 65
Omnibus, 3
omnidirectional sound, 139, 166
on line catalog, 126
on-line systems, Information America, 164
optical effects, caustic, 161
optical fibers, 53-58
optical systems, 9, 166
orbitals, 83-84, 166
orbiting instruments, 56
organisms, 166
ossicles (ear), 139
The Oz project, 169

P-Q

Pacific Data Images (PDI), 4, 31-32, 166
particle systems, 64, 166
PCs (personal computers), 36-38, 130-134
PDI (Pacific Data Images), 4, 31-32 166,
Peachey, Darwyn, 84
pencils, 89, 166
penumbras, 90-91
performance animation software, 166
Perlin, Ken, 90-91
personal assistant agents, 19
perturbed normals, 91-92
phase, chaos, 151
phase information (sound), 143
phenomena, natural (simulating), 63-64
photochromic glass, 50, 167
Photomorph (North Coast Software), 38
piano lessons software, 130
Pickover, Clifford, 110
piezoelectric, 167
  materials, 54-58
Pines, Josh, 90-91
pinna (ear), 139, 167
pixels, 167
plans, 167
point attractors, 150, 167
Polhemus' FastTrack device, 13
portable game systems, 129-130
portable game units, 167
PowerGlove (Mattel), 76
precise values, 118
primitive ellipsoids, 87
primitives, geometric, 62, 163
probes, proximal, 167
programs, 16-17, 36-38, 160
projection Virtual Reality, 74, 77, 80, 167
projections, 167
proximal probes, 167
Prusinkiewicz, Przemyslaw, 109
psychoacoustic, 167
  research, 139
puppets, 162-169

quadraphonics, 167
quaternions, 107, 112-114, 162, 167

R

radiosity, 89, 167
randomness, 167
ray tracing, 66-67, 87-89, 113, 160, 167
real-time, 167
receivers, 167
recording, 138-139, 141-145
  see also sound
Reeves, William, 84-86
reflection, 167
reflection maps, 66, 167

Index

reflective
  surfaces, 66, 167
refraction, 67, 167
region of
  complexity, 149
relationships,
  class/action, 120
rendering, 65,
  82-84, 88-91, 167
Renderman (Pixar),
  23, 34
repetitive
  behavior, 150
reproduction
  (Game of Life), 95
resolution, Sega
  Game Gear, 130
retrieving,
  information, 21-22
retroreflective, 167
Return to Zork, 131
RHS (right-hand-side)
  actions, 121, 167
Riddle, Jay, 34
robots,
  nanoscale, 166
Rogers, Craig A.,
  51-53
Rohrer, Heinrich, 45
Rosendahl, Carl,
  12-13
rules, 94-95, 120-121,
  167-168

S

Scanlon, Susan, 7
scanning tunneling
  microscope (STM), 45,
  168
scenes, 3-D
  (lighting), 65
Schrödinger,
  Erwin, 42
Sega, 128-129
  Game Gear, 129
Sega-CD
  Virtual-VCR, 129
self-monitoring
  systems, 55-56
self-organizing
  artificial life, 168
self-repairing
  concrete, 50, 56-57
self-similarity, 106, 168

Sensorama, 73
sensors, 53-54, 166
services,
  interactive, 164
sets, 107-110, 114,
  118-119, 122-123, 168
shaded models, 168
shape-memory alloys,
  embedded, 162
  metals, 55
  models, 168
shapes
  3-D, digitizing,
    62-63
  geometric,
    creating, 63
  quaternion, 162
shareware
  programs, DMorph
  (Dave Mason), 36
Shinya, Mikio, 89-90
shopping, 126
SIGGRAPH, 4,
  29, 168
Sigmund, Karl, 98
signal
  processing, 168
SIGS (Special Interest
  Groups), 98-99
Silicon Graphics, 9, 13,
  34
SimGraphics, 10-14
SimLife, 100
Simulation Virtual
  Reality, 77, 80, 168
simulations, 65-66,
  84-88, 90-91, 163
simulator sickness, 80,
  168
simulators,
  flight, 163
situational
  awareness
  (sound), 145
smart materials,
  50, 168
  goals, 52-58
  requirements of, 51
  research centers, 52
  sensors/actuators,
    development,
    53-55
  structure of, 52

Smart Structures
  Research Institute
  (SSRI), 50
Smythe, Doug, 29-32
SoftImage, 13
software
  3-D graphics, 74
  antiviral, 103
  Cellpro program, 100
  Edutainment,
    130-131, 162
  El-Fish
    program, 100
  performance
    animation, 166
  ray tracing, 87-88
  SimLife, 100
  The Miracle, 130
  virtual reality, 78
  see also programs
solution
  chemistry, 44, 168
sonic imagery, 168
sound, 138-146, 160,
  166-169
  see also hearing;
  recording
soundwaves,
  139-140, 168
source (light), 168
space, 3-D, 107, 160
spatial viewing, 72
spatially enhanced
  stereo, 138, 168
speakers, stereo sound,
  139
Special Interest
  Groups (SIGS),
  98-99
speech recognition
  technology, 168
spiral fractals,
  106-108
spot light, 65, 168
SSRI (Smart Structures
  Research Institute), 50
stable system, 168
Starbase virtual
  reality entertainment
  centers, 79
statements, IF THEN,
  121
Stealth virus,
  103, 168
stereo, 138-140, 168

stereophonic, 168
stereoscopic glasses, 70, 73, 76
  CrystalEyes, 77
stereoscopy, 73-74, 169
STM (scanning tunneling microscope), 45, 168
Stoned virus, 102, 169
Stookey, Donald, 50
storylines, non-linear, 166
strange attractors, 150, 169
Strike Commander game, 132-134
Super Mario Brothers (Nintendo), 11
SuperCockpit, 75
SuperFluo, 5-9
surface tension, simulating, 87-88
surfaces, reflective, 66, 167
survival (Game of Life), 95
Sutherland, Ivan, 73
swells, 82
system controls, 119

T-U

Tapestry program, 153-157
techniques
  lighting, 65
  modeling, 62-65
    Lathe, 63
Terminator 2: Judgment Day, 35-36
texture maps, 65, 91, 169
theme parks, virtual reality, 80
THX sound, 169
tidal waves, 82, 169
time/phase difference, 140, 169
trackers, 169
trains (wave), 82
transaural processing, 142-143
transistors, video games, 127
trends, 169
Troubetskoy, Eugene, 87
truth value, 169
Turbo Express (NEC), 129
Turk, Greg, 106
TV commercials, 5-7
tweening, 28-29, 169

Udd, Eric, 55
universe of discourse, 120, 169

V

V-art, 73, 169
VActor/Performer system, 14
VActors (virtual actors), 10-12, 169
variables, fuzzy, 120, 123, 163
ventriloquism, 169
vibrations, monitoring, 56
video, live, 131
video games, 126-129, 130-135
  see also Edutainment; entertainment
video music, 8
video phones, 169
Videoplace, 74-75
Viewpoint, 11, 12
Vincent, Julian, 50
virtual actors, 19-21, 169
virtual camera, 169
Virtual Interface Environment Workstation, 75
Virtual Kitchen, 79
virtual puppets, 17, 169
virtual reality, 23, 70-80, 134, 162, 169
Virtual World Centers, 79
virtual worlds, 70-80, 169
viruses, 101-103, 160-161, 168-169, 170
viscosity, 170
Vitz, Frank, 3
von Neumann, John, 94
voomies, 80, 170
VPL Research, 75

W

Wada, Benjamin, 56
Wahrman, Michael, 4
Walczak, Diana, 3
waldos, 3, 170
Walker, Chris, 12
Walt Disney World, automatons, 148
Walters, Graham, 5
warping, 29, 32-33, 170
water, animation, 82-92, 170
Watson, Thomas J., 112
Watt, Mark, 85, 89
wave arrival time, 170
waves, 82-87, 90-91, 160, 168-169
Whitted, Turner, 84
Williams, Steve, 34
Willow, 32-33
Windows program, 36-38
wings (airplane), self-monitoring, 51-55
WinImages:Morph (Black Belt Software), 36
wireframes, 65, 170
witness points, 170
worms, 101-103
writing, music, 97

Z

the Z factor, 169
Zadeh, Lotfi, 118
Zimmerman, Tom, 75

Want to Know More About These Cutting Edge Topics?

Here are some other Sams Publishing titles to look for...

The Magic of Image Processing

Discover phenomenal new dimensions of image processing. This unique guide is intended for everyone who is affected daily by image processing. The book discusses the ethics of photo manipulation, examines how Hollywood and Madison Avenue use image processing, and includes an introduction to popular image editors such Aldus Photostyler for the IBM and Adobe Photoshop for the Macintosh. Learn how images are manipulated and why. The disk contains fully implemented source and shareware to edit images. 0-672-30315-9

Artificial Life Explorer's Kit

Bringing users to the forefront of the hot, new, fascinating field of artificial life! This exciting book presents topics in a simple, direct style that allows readers to browse and experiment on their own. This book covers the history, state of the art, and future directions of artificial life. It also includes a disk with demo programs and artificial life simulations. 0-672-30301-9

Virtual Reality and the Exploration of Cyberspace

Comprehensive coverage of virtual reality for the high tech enthusiast! This book/disk set provides a nuts-and-bolts introduction of this emerging technology. It includes a disk containing virtual reality shareware, demos, and commands for the DIASPAR Virtual Reality bulletin board. The book discusses the social, political, and business implications of virtual reality. 0-672-30361-2

Morphing Magic

This book provides the secrets behind this fascinating graphics technique. Morphing maniacs will get in-depth information on electronic mutating approaches, plus the problem associated with creating realistic rendered images. The book features detailed descriptions of the use of morphing by animators, educators, and Hollywood's special-effects gurus. The disk contains fully implemented source code so readers can start morphing images on their own PCs. 0-672-30320-5

Tricks of the Graphics Gurus

Expert tips for creating and using graphics! This detailed tutorial teaches readers about basic concepts, issues, and general algorithms and techniques for programming graphics. Topics include fractals, animation, modeling, 2-D and 3-D graphics, ray tracing, morphing, and image processing. The book includes two disks containing all the programs and files needed to run the graphics covered in the book. 0-672-30308-6

Add to Your Sams Library Today!

The easiest way to order is to pick up the phone and call

1-800-428-5331

between 9 AM and 5 PM EST, or you can write to us at

11711 North College Avenue, Suite 140, Carmel, Indiana 46032